GEOGRAPHICAL EXCURSIONS IN LONDON

Hugh J. Gayler

University Press of America, Inc.
Lanham • New York • London

Copyright © 1996 by
University Press of America,® Inc.
4720 Boston Way
Lanham, Maryland 20706

3 Henrietta Street
London, WC2E 8LU England

Library of Congress Cataloging-in-Publication Data

Gayler, Hugh J.
Geographical excursions in London / Hugh J. Gayler.
p.   cm.
Includes bibliographical references and index.
1. London (England)--Historical geography. 2. London (England)--
Tours. I. Title.
DA684.25.G39   1996   914.21'204859--dc20   96-13642 CIP

ISBN 0-7618-0327-0 (cloth: alk. ppr.)
ISBN 0-7618-0328-9 (pbk: alk. ppr.)

Contents

Figures

Plates

Preface

This book is the result of trying, and failing, to find a geographical guide to London. I was looking for the present-day equivalent of the *Guide to London Excursions* that was prepared for the 20th International Geographical Congress in London in 1964 (Clayton, 1964). There are, however, many excellent guides and walking tours of London's history and major architectural features, as well as short walks of specific geographic areas, such as the legal quarter, Jack the Ripper's London or Little Venice. There are any number of tourist guides - good, bad and indifferent - of the what-to-do, where-to-stay variety. But none of these guides fulfills the purpose of understanding London's urban development, especially from the point of view of what is happening, and why, on the ground today. Furthermore, much of the guide material relates to central London, while most areas outside of this, which have played a significant role in the growth and development of the capital, are hardly written about and rarely visited.

Since I was charged with taking a group of university students on a two-week field course to London, and students like a text book that is relevant to the course at hand, it was incumbent on me to come up with some sort of guide. Moreover, I certainly wanted to be organized about what I was doing and to make sure that the field excursions were relevant and feasible. Thus, I spent a busy few months putting together the first draft of what you see here; and those first students, to whom I shall always be grateful, suffered a great deal as the guinea pigs, when walks turned out to be far too long, bus routes had mysteriously changed, and various strikes meant whole days had to be switched around.

A second draft of this book was prepared and I was then encouraged by a number of fellow geographers to seek a publisher. Thousands of students visit London every year, and it was thought likely that such a text would be very marketable. Also, many other visitors to London, who readily buy historical and architectural guide materials, may be interested in a urban geography of the city. This would be set up as a series of urban issues, past and present, that can be appreciated in the landscape and organized as walking tours and excursions by public transportation in different parts of London.

In preparing the various walks and excursions, I was helped by an incredible list of reference materials. So much so, that there should have been no possibility of my getting it wrong. If I have, the fault is mine alone. I have not included a voluminous bibliography; instead, I have recommended further reading in each chapter, and these are listed in a bibliography at the end of the book. Many of these works are in fact texts on London and include a large bibliography in their own right. I refer specifically to the following: Clout and Wood, 1986; Hall, 1989; Hoggart and Green, 1991; Porter, 1994; and Thornley, 1992.

The structure of the book is a series of chapters, outlining either a walking tour of part of central London or a combined public transportation and walking tour of more distant suburban areas. Each of the nine chapters involves a day or part of a day excursion, although I hesitate to give actual hours; the field course has now run on two occasions, and in some cases the same route has been covered in two very different times! Much depends on walking speeds, length of stops (including lunch), times waiting for public transportation, side trips of interest and time spent talking to people (both of which simply happen and cannot be planned in advance), and the weather. It is not a matter of life or death to go by the book completely. Individual discretion can certainly be used, for example, to forego part of the route or carry over some feature to the next day.

The nine chapters focus on particular areas of London and thus involve a certain time frame in terms of the development, and redevelopment, of the city. In some instances, especially in central London, that time frame can be exceedingly long and there is much to comprehend in a relatively short walk. Often, there are a number of parallel issues; an issue may be covered at one point on the tour, but the reader should be aware that more examples of the same thing may be seen later in the day, or even on another day.

The urban geographer is being exposed in London to a major world city in economic terms, and the various institutions and functions associated with this are outlined, in particular finance in the City of London itself. In the British context, London is very much a primate city and the focus of a wide range of national and international functions, including the head-office role, high-order retailing, higher education, medicine, the law, information and media, entertainment and tourism. Its long history as the capital of a nation state, and a state that heralded the Industrial Revolution and was for long the major world power, has inevitably led to London having an important governmental and cultural role. This will be seen in the City of Westminster and other parts of the central area. London was the world's largest port, and the dying embers of this will be seen amid the glitter of new office blocks and yuppie housing. Greater London is home to over six million people, and our excursions in both inner and outer-suburban areas will examine these residential landscapes, including variations by period, class, and type of housing, as well as the nature of the community itself.

Since this is a field course, there is an opportunity to experience the built environment or urban fabric - the manifestations of the present and past generations in urban form, streetscapes, and building designs, and how the various activities interrelate with those. I draw attention to this throughout the book. We can also see how Londoners carry on their daily lives, and I encourage the reader to talk to local people. It is not too difficult since everyone will find him/herself in a shop, restaurant, bar, hotel, theatre, bus or wherever the public are to be found. Inevitably, a group of students will attract attention, especially in some residential areas, and it may be a local person dying of curiosity who will start up the conversation.

The book is designed to be either instructor-led or used by the individual. Each chapter has directions to be taken on foot or by public transportation, and there are maps indicating the route. It is always advisable to have a pocket-size *London A-Z Street Atlas and Index*, just in case one gets lost or wishes to deviate from the route. For ease of using public transportation, it is worth purchasing a London Travel Card for the appropriate zone that will be reached. This allows for unlimited travel by bus, London Underground and British Rail within the zonal limit for whichever time period. Note that a day travel card cannot be used Monday to Friday before 09.30, while a weekly card requires a passport photograph at the time of purchase. The four days in central London (Chapters 2-5) are solely walking tours, although public transportation to and or from the particular area may be necessary depending on where one is staying. Only one excursion, the final leg of the East End transect, involves travel beyond the limit of the London Travel Card. In this instance, a British Rail return ticket (or group ticket if there are more than 10 people) can be purchased beforehand to cover the journey from the Zone 6 limit at Upminster to Basildon. Directions in the text include route numbers for London's buses, but remember these are subject to change, as I have on occasion found out. You would do well to check the information when you travel; London Regional Transport can give advice and supply bus route maps.

These excursions in London purposely do not include any travel by private bus transportation. Apart from the considerable expense, such transportation is frequently impractical or unnecessary, and slower than using the railway. Also, one is removed from the sights and sounds and people of London by a layer of glass (worse still, tinted glass!). Absent also are visits to various tourist and cultural features; we pass many of them on our central London walks, and refer to them in the context of London's development, but we do not go in. The integrity of the excursion could be upset, and the time taken up could easily wreck it! However, time should be set aside so that you can return at your leisure to the feature of your choice. The two-week field course had three free days and all evenings free in order to do the things that all visitors, including me, may want to do.

The chapters are selective. Instructors and students who are staying longer can extend the tours or incorporate more activities. For example, the walking tour of the City of London could walk through the Museum of London. Those staying for less than the nine days can cut and paste their own excursions to fit the objectives of the particular field study. In being selective you will see that, apart from Greenwich, the excursions are all north of the river. No slight was ever intended on folk who live south of the river; the fault lay perhaps with my poorer mental map of that area, resulting (as others have also noted) from the brilliant conception of the London Underground map (most lines are north of the river) versus the chaotic display (and involved geography) of the British Rail map for the area south of the river, plus the fact I have only ever lived north of the river!

Once outside the central area, it is possible to replicate much of London's urban geography in its various inner and outer suburbs. Instead of Metroland in the inter-war suburbs of north-west London, I could just as well have chosen Ideal Homes, and its definitive architectural heritage, in London's south-east suburbs. Likewise the transect from Shoreditch to Basildon could have been Southwark to Crawley. The space-time model can be moved and still incorporate the major features of London's growth and development; this book is thus a challenge for someone to bring on line areas that he/she knows well.

To have the opportunity to take students around London must be viewed as a labour of love. London is an incredibly fascinating place, in spite of the dirt, noise, poverty, congestion and run-down state of some of the infrastructure, and every visit brings new discoveries, changes and experiences which I attempt to share with students. A field course of this nature does not have to be expensive. Reasonable university and college bed and breakfast accommodation, centrally located, is available in the vacation; walking and travel cards come cheap; and students quickly find the good and cheap eating establishments off the tourist beat.

It is important as human geographers to get beyond the classroom, the library and the armchair. Field courses are a lot of work, but the rewards are great as we see our discipline and its concepts applied and in action. Good luck with the book and testing it in the field.

Brock University   Hugh J. Gayler
Ontario, Canada

December 1995

Acknowledgments

This book would never have been possible without the considerable assistance and professional judgment of a number of people. First, I would like to thank Professor Peter Hall, Bartlett School of Planning, University College London, for reading and commenting on an earlier draft of the book, and also to the various anonymous reviewers. My colleagues at Brock University gave freely of their time in talking about various issues relating not only to London but also carrying out work in the field, and as always I am very grateful to them. I would like to thank Dr. Will Webster, Dean of Social Sciences at Brock University, for his continued support, and especially for the financial contribution towards publishing this book.

The evaluations of my students were very important in assessing the success of the field course and its particular excursions. I am indebted to the following who were the guinea pigs on that first course: Dale Andrews, Shelley Bannon, Jennifer Brahaney, Sean Brathwaite, Glenn Cook, Tracy Hurd, Mark James, Greg Johnston, Darryl Koop, Tammy Naughton, Shelley Partridge, Pauline Scoffield, Bruce Steinburg, Deanna Walton, Scott Watson and Christina Winkworth. Thanks also to the following who went on the second course: Barbara Boyes, Brian Furry, Jon Gemella, Sheila Gillies, Julia Hessey, Tara Johnson, John Mackenzie, Keith Sikkema, Dino Varas and Janice Zanon.

Translating a student guide into the book you see involved three of my colleagues in a very direct way, and I would like to thank them here. Loris Gasparotto, our departmental cartographer, drew all the maps, maintaining his usual high standard amid the pressures of a busy office and the frustration of interpreting my sketches. Colleen Catling, our departmental secretary, was a frequent saviour when I ran into difficulty preparing camera-ready copy. Divino Mucciante, the university's photographer, prepared the illustrations for publication, and I thank him for all his efforts in this regard.

I would like to recognise a number of people or institutions who helped me with written materials or illustrations, including the Barbican Centre, the Crown Estate, John Jackson (Professor Emeritus, Brock University), the London Boroughs of Barking & Dagenham, Barnet, Camden, and Hillingdon, Tony O'Regan (Head of Media & Public Relations, London Borough of Tower Hamlets), and Peter Wade (Head of Public Affairs, Canary Wharf Ltd.).

Eileen Martin and Michael Gould were (and are!) long suffering friends who were dragged against their will around London to investigate another piece of the puzzle for this book, and who offered valuable advice in the process. My children, Jonathan and Emma, have graciously, and at times not so graciously, suffered the same fate. My love and appreciation to them also.

Finally, my thanks to the publishers, University Press of America, and to Michelle Harris (Acquisitions Editor) and Helen Hudson (Production Editor) for making this book possible. As a Canadian writing on a British subject for an American publisher, I apologise in advance for adopting Canadian rather than American spelling, and using phrases and words from both sides of the Atlantic Ocean.

Chapter 1

Introduction:
London as an Urban Study

London, along with New York and Tokyo, is a premier world-class city. It has long ceased to be the world's largest city; but its central area has one of the largest concentrations of jobs in the world, and ones that play a pivotal role in the global economy. The City of London, that small, square-mile area at the heart of the capital, is the financial power house, focusing on banking, stockbroking and insurance. Nearby in the City of Westminster is the central government for Britain and a range of administrative offices for the Commonwealth, foreign governments and other world bodies. This area has a disproportionate share of the head offices of British-based, trans-national corporations, as well as offices and institutions associated with commerce, medicine, education, law, arts and entertainment, religion, media and information technology. London's West End is the largest and most important shopping district in the country. Britain is a constitutional monarchy, and the royal family, the Court and a longstanding social elite reside in London or maintain a London residence. London's history, culture and royal pageantry result in it being the most important tourist destination in the country. It is perhaps ironic that while Britain's position as a world power has declined during the 20th century, this has not been allowed to affect London's world-class status.

As we examine the various parts of London, whether the central area or its suburbs, we will become aware of how central this city is to an understanding of the British economy, society, government and psyche. It was a role that began nearly two thousand years ago and was reinforced as Britain became a nation-state, a mercantile and colonial power and an industrial state. Changes to transportation and information technology and the centralizing tendencies of government and business in the recent past have continued London's primate-city role within Britain, as well as its growing world-city role. Nearly 12 million people, one fifth of the country's population, live within the Metropolitan area, while nearly 18 million, one quarter of the population, live within the South East (Figure 1.1).

London's pivotal role in so many aspects of British life causes great consternation in other cities and other parts of Britain, contributing to a resentment about Londoners and the South East. Either London has the best that Britain has to offer, or its overwhelming media presence makes sure that London is promoted over its provincial counterparts. So much so that when overseas visitors think of Britain and British institutions, the examples used are all too often from London: tennis is synonymous with Wimbledon, opera with Covent Garden, royalty with Buckingham Palace, exclusive shopping with Harrod's, air travel to anywhere from Heathrow, and on and on the list can go. Billions of pounds of decentralization measures later and what lies beyond London are relatively unknown yet fascinating towns and cities, declining industrial areas, and mile upon mile of beautifully manicured countryside, villages and wild mountain areas. A few cities, such as Oxford, Cambridge, Stratford, York and Bath, stand out from the rest for obvious historic and cultural reasons, while certain rural and mountainous areas are favoured more than others, all of which are accompanied by peak-season pressures on the environment and service facilities. But London remains the most talked about, written about and frequented part of Britain.

A definition of London

Whilst London is central to British life, and certainly to tourism, it is a place that is hard to define; and once one has done so it is even harder to comprehend for it is so diverse and so disorganized. First of all there is *no* London! There is the anachronistic City of London from Roman and medieval times, now containing fewer than 10,000 people. Then there is Greater London which came into existence by Act of Parliament in 1964, consisting of a two-tier system of county and 32 London boroughs, but excludes the City of London (Figures 1.1 and 1.2). This area of roughly 6.5 million people today took over the role of the County of London and its boroughs (roughly co-terminus with Inner London on Figure 1.1) that had existed since 1889. Greater London also absorbed the whole of the county of Middlesex, plus the adjacent, urban parts of Kent, Surrey, Hertfordshire and Essex. However, Greater London as metropolitan county government was

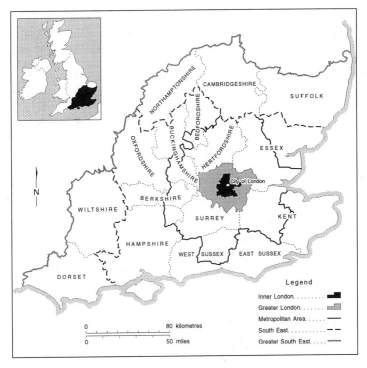

Figure 1.1 London and the South East defined

abolished in 1986 (together with all the other metropolitan counties in England), and London is now run by its 32 boroughs and the City of London, a myriad central government organizations (for example, the police, public transport, and roads) or combinations of local government.

Since Greater London roughly encompassed the 1939 built-up area, it has been necessary for various administrative, statistical and planning purposes to define a larger area than Greater London: hence the Metropolitan Area which extends almost 40 miles from central London (and including London's post-war growth) and the South East Region beyond that, the two areas together constituting ROSE, Rest of the South East or Roseland. Growth even further out (for example, the Mark 2 New Towns of Northampton and Peterborough) has resulted in the definition Greater South East, and London's influence is even reaching beyond that as a result of long-distance commuting.

London's spatial diversity

The geography of London is remarkably diverse. We shall see that its various functions are highly concentrated spatially. The City (short for the City of London) is finance, the southern part of the City of Westminster is government, the central part (the West End) is commerce, retailing and entertainment (again segregated by function), lower Thameside is what remains of heavy industry.

There is a very clear age difference as one proceeds out from the centre, and the effect this has had on physical forms and architectural styles makes for considerable contrasts between the Roman and medieval city, the post-1600 West End, the ring of early railway and industrial suburbs, the late 19th-century by-law housing areas, the outer, inter-war, semi-detached suburbia and the expanded provincial towns and New Towns beyond the Green Belt. Over the years, and especially since the Second World War, most of these older areas have been subject to considerable redevelopment.

There is also very considerable social variation in London. After 1600 upper and middle-class London was focused on the West End and a growing ring of suburbs about this stretching to the north, west and south-west. Meanwhile, working-class London was heavily focused on the East End, inner south London, lower Thameside and the Lea Valley with pockets in other industrial areas such as Willesden-Acton in West London. As a result, London is a very politicized city. Voting patterns and political control (by the Labour or Conservative parties for the most part) are correlated with social class. Also, since the 1930s, the county and most of the boroughs have principally been controlled by Labour, while central government at Westminster has largely been in Conservative hands.

There are also some stunning contrasts in London with some of Britain's most poverty-stricken areas adjacent to some of its richest, urban fabric on the point of collapse close by some of the best maintained, a well-paid white-collar workforce oblivious to the homeless sleeping in doorways, and private wealth pitted against an increasingly under-funded public infrastructure.

This social imprint, however, is by no means static. We will see middle class areas in West London that have deteriorated and been increasingly occupied by a poorer population, working-class areas such as Docklands that have been yuppified, areas such as Stepney that have gone from almost all White to heavily non-White, and public-housing areas where housing has been sold and gone up-market.

Figure 1.2 Greater London and the London Boroughs

The growth of London

London's growth has been mostly slow, incremental and highly radial along its major roads, rivers and railways. Small communities grew up which focused on a particular estate development at one point in time, around a medieval village or a 19th-century dock or factory, or later around a station on the expanding suburban railway or Underground (subway) network. London was disorganized in favour of very local communities where people lived and worked and interacted; and if they ventured out, it was to work, shop and seek various forms of entertainment in the central area. So dependent were Londoners on the central area, that strong regional centres in the suburbs have been lacking until fairly recently, particularly in the older, inner suburbs. Meanwhile, the insularity of the various areas in London leads to the old adage: never ask a Londoner the way to anywhere, if in doubt ask a police constable or a taxi-driver.

Londoners feel themselves to be part of these small communities, and the reorganization of government into 32 large boroughs (plus the City of London) has done little to change this feeling. Most of these boroughs can be described as little more than marriages of convenience, designed to improve the provision of local services. Most have no single focus or one commercial core; even local government services can be split between the many former town halls. Open rivalry between different wards, former boroughs or political groups is still evident. Many of the boroughs could not agree on a new name, resulting in compromises, as in the historic name of Havering, or combining the former names, as in Kensington and Chelsea. There was probably greater loyalty to the County of London, and later Greater London, both of which have been abolished.

The enormity and diversity of London result in it being a difficult city to come to terms with. Most people, tourists and Londoners alike, do not venture much beyond its central area, an area bounded by the City in the east, the River Thames, Westminster, Victoria, northwards to Oxford Street and east again to the City. Obviously, places where people live or stay and certain specialized areas such as the South Kensington museums, the South Bank, Little Venice, Kings Road and Docklands will take people outside this central area. Also, expanding work places have widened what constitutes the central area towards Paddington, Kensington, Kings Cross, south of the river, and Docklands. But much of London remains largely unknown and rarely visited, especially its predominantly residential suburbs.

London day excursions

Our excursions will take account of this diversity in terms of age (and distance from the centre), form, function, social composition, type of housing and the nature of urban regeneration In the course of doing so we will look at the many problems that London faces in being a world city, the capital of Britain and a major metropolis.

The first area to be examined is the financial centre of the City of London; and it is preferable to walk its maze of narrow medieval streets and see its working population on a weekday. The major institutions are situated amid extensive redevelopment that has had to replace bombed and worn-out buildings and accommodate various technological changes. Contrasting with this is London's later, second city of

Westminster, the hub of its administrative and executive government institutions, as well as the head-of-state functions and some of Britain's most prestigious residential areas. A third excursion will be in the northern and central parts of the City of Westminster, including the Regent's Park and West End areas, where there is an incredible mix of commercial functions beyond those of government and finance, as well as London's first suburbs. A final, short central-area excursion will focus on the inner-London suburbs of Bloomsbury and St. Pancras, developed in the 18th and 19th centuries, and including the University quarter and the early railway developments along the Euston Road.

The next excursions will focus on suburban London. The first two days will involve us in a transect of London's East End, commencing in the Shoreditch-Whitechapel area to the north-east of the City of London, proceeding through Stepney and Poplar to the turn-of-the-century suburbs of West Ham and East Ham. The inner areas constituted some of London's worse slums, and since 1900 and especially following extensive wartime destruction, much of the area has been replaced by public housing. Further out there has been less renewal and more refurbishment of the existing housing. The collapse of the old order, especially the decline of unskilled and semi-skilled jobs in industry and transportation, has resulted in considerable deprivation in many of these areas.

The following day will focus on post-1920 East London. An important aspect to slum clearance, urban renewal and accommodating population growth has been the planned and unplanned migration of people to the urban fringe. We will first view the inter-war public housing estate at Dagenham (the largest ever in Britain) and then travel beyond the 1938 Green Belt (and the boundary of Greater London) to examine post-war developments, in particular one of London's New Towns at Basildon in Essex.

A third day in London's East End will encompass the very contrasting environments of the Docklands, one of the world's largest urban regeneration projects where the controversial principles of Thatcher's Britain were put into practice in the 1980s. Part of this area, the office complex at Canary Wharf, has been dubbed London's Third City. Meanwhile, million-dollar condos rub shoulders with some of Britain's most rundown housing.

The next excursion will take us to the north-west London suburbs to examine the extensive area of low-density, private housing that was built in the period after the First World War. Unceremoniously dubbed Metroland, after the major developer, or Semi-Detached London, after the principal form of housing, these areas have withstood well the various negative aspects of urban change.

The final excursion, to Hampstead and its environs, includes an eclectic mix of features, but ones of great significance in London's development. Hampstead is one of London's urban villages; and some of the characteristics of the medieval village and the later spa settlement near the mineral springs can be seen amid the late 19th century suburban growth. Nearby, Hampstead Heath, Golders Green and Hampstead Garden Suburb warrant exploration.

These various excursions are only a beginning to discovering the wealth of London's urban geography. However, this book is intended to provide a basic geographic framework, setting into context the major periods of development, examining various aspects of form, function, institutions and peoples and looking at the many problems associated with living in London today. In no way do these excursions cover all of London, but it is hoped that the reader will be able to use these materials to explore and be able to appreciate the geography of other parts of the Metropolitan area.

Chapter 2

City of London
(The City)

The City of London is approximately one square-mile in extent and is situated on the north bank of the River Thames in the heart of today's capital. It contains the original Roman townsite, dating back to the 1st century A.D., which was delimited by a wall and various gates, and also includes some of the later medieval urban development, particularly to the west along Fleet Street (Figure 2.1). The site chosen for the City of London probably reflects the first place upstream from the North Sea where the river was not flanked by extensive marshlands; and thus it was possible to build a settlement free from regular flooding and on other than alluvial material. Also, the river could be more easily bridged here than anywhere else downstream.

The City became the financial capital of England and the centre for many of its legal institutions, commercial undertakings, port and maritime operations, and craft and professional associations (guilds). In the medieval period it obtained for itself considerable wealth and a measure of independence from the Crown (and later the central government) which still features in its various institutions. Britain became a major colonial and world trading power after the 16th century and after the 18th century the leading country of the Industrial Revolution. Whilst these led to the growth of London well beyond the City's limits, increased trade and industry were responsible for the City gaining even more wealth and influence. The more recent shift to a service based economy has likewise benefitted the area.

By 1600 the London area had emerged as the largest world city, with a population of over 200,000, a position that was maintained until well into the 20th century. By 1800, and the first official census, the

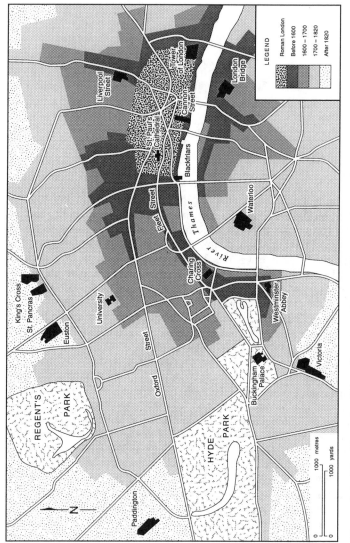

Figure 2.1 The Growth of London since Roman Times (After Hall, 1964)

population had reached nearly 1 million, a century later it stood at over 4 million. From a pre-war peak of 8 million, the population has now fallen to 6.5 million, and London is far from being the largest city in the world. However, along with New York and Tokyo, London has become a financial power house and a first rank world-class city. The City of London is the financial centre; its international banking, stockbroking, insurance and other financial roles link the global economy to the various European regional and local centres.

The City's other functions have waxed, and usually waned, over the years. It has retained its important legal status, including the highest courts in the land (outside of the House of Lords and the highest court of appeal), and the various Inns of Court. With so many historical and cultural sites, the City is also an important tourist destination. Its history, its institutions such as the various guilds, and its wealth and influence have helped the City to maintain a measure of independence and political power out of all keeping with its population size. From a high of 127,000 in 1851, the City's population had fallen to 4,800 by 1961; it has doubled in population since then as a result of redevelopment schemes. Neighbouring communities of much larger size have long since been amalgamated, but the City has successfully resisted such incursions. It was never part of the administrative County of London after 1889, nor Greater London after 1965, and it has retained its archaic and far from democratic form of government and many of its own institutions (for example, police). Now that Greater London government has been abolished, the 32 London boroughs can join the City on a near equal footing!

Those functions that have waned, in addition to the residential one, include industry, warehousing and the distributive trades, especially those related to maritime trade. The City was synonymous with the Port of London until late medieval times when increased traffic resulted in more riverside wharves downstream. London Bridge marked the upstream limit for ocean-going vessels, and goods were transferred at this point to or from barges (or lighters). Since the 1970s this activity has all but collapsed. Likewise, the City was an important wholesale food centre for fresh produce with a number of major markets; these have either gone or are going to better locations in the London area. Finally, it was the centre for the printing and publishing industry and contained all of Britain major daily newspapers; again, these have almost all gone to other locations.

The City of London is very much an office employment centre. By day (or rather Monday to Friday) its 10,000 population expands to nearly 300,000, part of the 1.1 million commuters who arrive in central London each day from the suburbs and provincial towns as far away as

100 miles. Approximately (and fortunately!) 85 per cent of these commuters arrive by public transportation. The City is served by eight London Underground lines; British Rail services into Liverpool St., Fenchurch St., Moorgate, Cannon St., London Bridge and Blackfriars Stations; British Rail's Thameslink through service; even a waterbus service to Tower Pier from Greenwich and Docklands has been tried; and a myriad bus services both local and long distance. Traffic is heavily focused into morning and evening rush hours, Monday to Friday. On the weekends, streets and public transportation can be quite deserted and most services are closed.

As small as it is, the City is spatially differentiated. First, there is physically a western and eastern city with two hills of gravel approximately 40 ft. above the river, separated by a small stream, Walbrook, now underground. The top of Ludgate Hill in the west and Cornhill in the east are the sites of St. Paul's Cathedral and the Bank of England respectively. Second, there are functional differences within the City. For example, the area to the east of the Bank is London's financial quarter, while at the western edge of the City is the legal quarter; and along the River Thames the remaining maritime activities can be found. The City has constantly been redeveloped and few buildings before the 19th century remain. However, the City still has an abiding medieval feel to it because the street pattern, the legal division of land and building lines still largely date from this period. There has been little largescale redevelopment until quite recently.

Our walk in the City of London will enable us to trace the major developments in this Roman, medieval, and more recent city (Figures 2.2-2.3). The major functions of the City, past and present, will be identified, as well as its important institutions, and various modern planning and development issues will be examined.

(Circle or District Line to Tower Hill Underground Station; exit and stop by the Roman Wall)

Roman London

Much of what remains of Roman London lies underneath the present city or is represented in the modern street pattern, but in the Tower Hill area, immediately outside Tower Hill Station, can be seen one of the fortifications at the eastern end of the wall around the Roman city. This walled town covered approximately 330 acres (the largest in Britain) was built about A.D.100 on what is now the City (Figure 2.1). The wall is evident in a number of places and the street pattern today can be seen to follow its course, especially on the north and east side (London

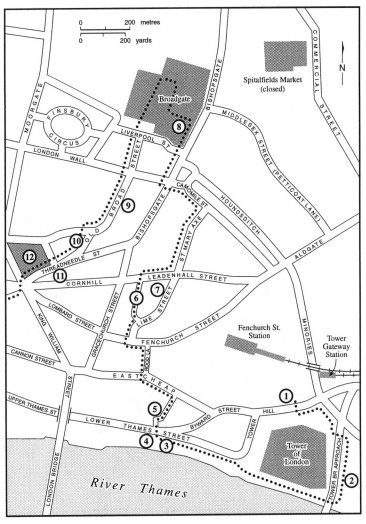

Figure 2.2 City of London Walk

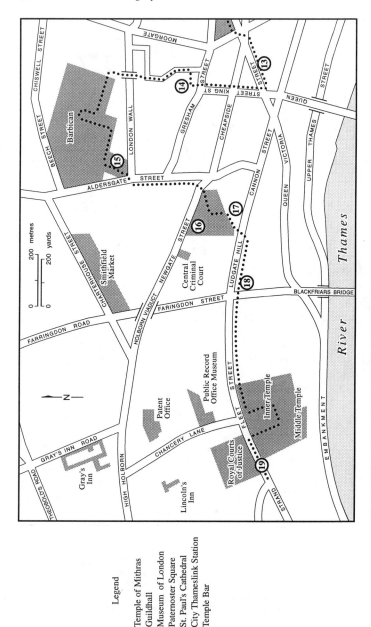

Figure 2.3 City of London Walk (cont.)

Legend

13 Temple of Mithras
14 Guildhall
15 Museum of London
16 Paternoster Square
17 St. Paul's Cathedral
18 City Thameslink Station
19 Temple Bar

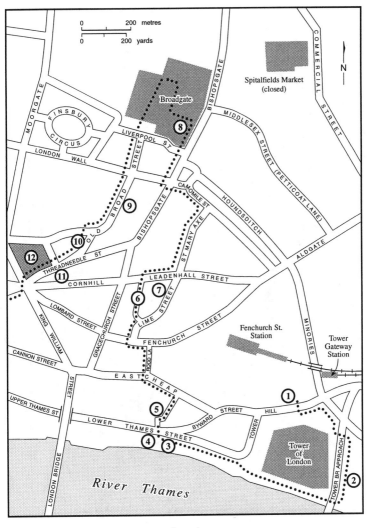

Figure 2.2 City of London Walk

Legend

1	Tower Hill Station	**7**	Lloyd's of London
2	St. Katherine's Dock	**8**	Liverpool Street Station
3	Custom House	**9**	National Westminster Tower
4	Billingsgate Market (closed)	**10**	Stock Exchange
5	St. Dunstan's Church	**11**	Royal Exchange
6	Leadenhall Market	**12**	Bank of England

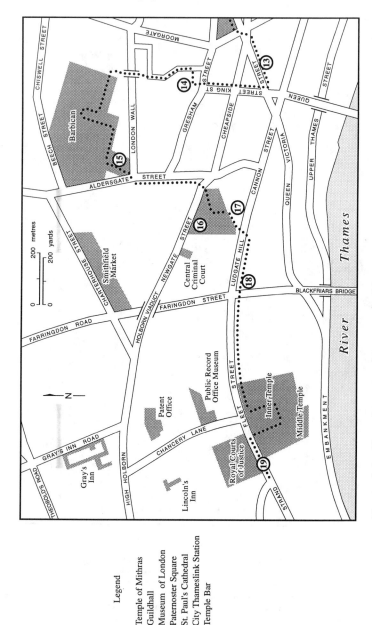

Figure 2.3 City of London Walk (cont.)

Legend

13 Temple of Mithras
14 Guildhall
15 Museum of London
16 Paternoster Square
17 St. Paul's Cathedral
18 City Thameslink Station
19 Temple Bar

Wall, Houndsditch and the Minories). The various gates through the wall are long gone but are evident in the street names; working anti-clockwise, these are Aldgate, Bishopsgate, Moorgate, Cripplegate, Aldersgate, Newgate and Ludgate. The Roman road system through Britain focuses on the City via these gates in spokes-of-the-wheel fashion. However, there is little evidence of the street plan within the walls; the present streets focusing on the Bank are probably medieval in origin.

On our walk through the City we will see other evidence of Roman London, including walls in the Barbican area and the Temple of Mithras in Queen Victoria Street. This is perhaps the most impressive Roman find in London, discovered in 1956 during excavations to locate the bed of Walbrook.

(Turn left along Tower Hill towards Tower Bridge Approach Road, cross the road and turn left into St. Katharine's Dock, walk through the Dock towards the River Thames and Tower Bridge, turn right along the riverside walk, under Tower Bridge and along the front of the Tower)

Tower Hill

This is one of London's most important historic and cultural areas and a major tourist attraction. Our walk in fact begins just outside the City of London in the former Borough of Stepney, now the London Borough of Tower Hamlets. The emphasis in this area are functions once associated in some way with the River Thames:

St. Katherine's Dock

This was the fourth of London's great 19th century and early 20th century docks, built by Thomas Telford and opened in 1828. The tidal nature of the river and the difficulties this posed to port facilities, as well as wharf and river congestion, were overcome by an extensive dock system with controlling lock gates. Hundreds of acres of alluvial floodplain, proximity to the City of London and little urban development at the time were important factors in this eastward spread of docks, warehouses and industry. (St. Katherine's Dock, however, was an exception; a community was destroyed in order to put in this particular dock.) St. Katherine's Dock was closed in 1973, and since then the warehouses have been converted (although too many have been demolished), a trade centre has been built, a yacht basin opened and other tourist and residential functions have been developed.

Although predating the London Docklands Development Corporation (LDDC), it is now part of its area and mandate.

Tower Bridge

Until the recent opening of the Queen Elizabeth Bridge on London's Orbital M25 Motorway, this was the last of the road bridges going downstream and the last to be built (completed in 1894). The bridge and approach roads are part of an inner area ring-road and at times can be quite congested. This somewhat detracts from it being without doubt the most magnificent of the London bridges. Indeed when an American company bought the old London Bridge to resurrect in a desert resort in California, they actually believed at first that it was Tower Bridge that was for sale! Because of the decline in river traffic the drawbridge is not often opened. The bridge was recently refurbished, and after being closed for decades because of the suicide risk, the upper walkways of the bridge have been enclosed and are now open to visitors.

The Pool of London and London Bridge City

This part of the river between Tower Bridge and London Bridge is as far as large ships can go. Hay's Wharf once bustled with activity as ships unloaded or transferred their cargoes to lighters (barges), but it is now part of the LDDC area; the dock has been filled in and the warehouses have been converted into offices and shops (Hay's Galleria). Downstream on the south side of the river is Butler's Wharf, a similar conversion. The local area, now called London Bridge City, contains many new office blocks and apartments and sites awaiting development. A major British Rail station serves the area, and connections to central London and Docklands will be improved with the completion of the London Underground's Jubilee line. HMS Belfast, the last of the World War II cruisers, is a naval museum and is permanently moored in the Pool of London.

Tower of London

This is one of London's prime tourist venues, but amid the history and the glitter it is a working army base, containing one of the Guards Regiments of the British Army - to look after the Crown Jewels, the Royal Armouries, as well as other 'defence of the realm' duties, many of which are ceremonial. There has been a fortress here since Roman times, but much of what you see dates from the Norman Conquest (when William I invaded England and defeated the Saxons in 1066). The central White Tower is a fortress within a fortress and is now used as a museum (Plate 2.1); it contains a chapel which is perhaps the best example of Norman church architecture in Britain. For centuries the Tower was also a royal palace, with residences for the various nobility, and one of the most notorious of prisons. For nobility and prisoners alike the river was an important means of reaching the Tower. For the last 150 years Tower Pier has been the London terminal for pleasure boat traffic to downstream and coastal resorts.

Medieval and Mercantile London

In the 1200 years between the Romans abandoning Britain in A.D. 410 and 1600, there was little physical growth of the City (Figure 2.1). What did occur was principally to the west (and now within the City) and across the river in the Borough area (the present London Borough of Southwark) where London's third and little known cathedral is to be found. Outside of that was mostly ribbon development along the river or along the Roman Fleet Street and Strand towards Westminster.

The City as the commercial capital was part of a functional diptych or duality, with the City of Westminster being the seat of government for England and the royal and political capital. In addition to commerce, the City of London contained important monastic orders; and their work can be seen in the Church, hospitals, schools and housing, especially St. Paul's Cathedral (founded in the 7th century) and St. Bartholomew's Hospital (founded in 1123). One of the monastic orders, the Knights Templar (founded in 1185) was colonized by lawyers a hundred years later and the Temple became the nucleus of London's legal quarter - the Inns of Court. The street pattern seen today dates from this period, and major catastrophes such as the Great Fire of London in 1666 and the Blitz of 1940-1941 did not result in much in the way of change to the narrow, twisting streets that evolved out of the Roman plan.

In the City itself most of the major buildings date from the period after 1600 when Britain became a major world trading and colonial power, and in particular after the Great Fire of 1666 when an area of approximately 440 acres was largely destroyed. There were ambitious plans to reconstruct London after this in the manner of continental town planning with broad, straight streets leading to and from squares and circles. The plans of Sir Christopher Wren, who was responsible for the present St. Paul's Cathedral and numerous City churches, were but one example in the late 1600s. However, the problems of amassing properties to allow for change resulted in the City Corporation

Plate 2.1 Tower of London from the River Thames, showing the White
Tower and Traitors' Gate in front (Photo: H.J. Gayler)

abandoning any such attempts, and the City continued to develop
piecemeal.

Part of the piecemeal development was the continued intensification
of development in the City. Until the 1850s, this population growth in
central London was a reflection of little or no public transportation and
little opportunity to decentralize residential development. The long
gardens of the merchants' houses were gradually built upon with
interior courts and narrow alleyways leading to the street. It was only
with the Blitz and post-war urban renewal that most of these courts
have disappeared, but it is interesting to note that largescale change,
especially regarding the road pattern, has not taken place. In spite of
new and taller buildings, widened roads and modern transportation, the
City of London is still seen as a medieval place. Our field excursion
after the Tower will encompass these medieval and later institutions:

(From the Tower turn left on Lower Thames Street)

Markets

Lower Thames Street shows the area's former maritime function with the Custom House and the now closed Billingsgate (fish) Market on the riverside. The City was once the wholesale and retail centre for the London area. Cheapside to the west of the Bank of England was formerly a retail street. (Cheap comes from the Saxon word to barter.) As the City's population has declined since 1851 so has retailing; the small amount that exists is associated with the lunch-time office trade or the specialists trades and professions to be found in the City (for example, the shops of the silversmiths in the Chancery Lane area, or the various legal book shops, gowns and wig makers for barristers).

A number of major wholesale and retail markets were established in or around the City, but traffic pressures and inadequate space for expansion have resulted in most of the markets relocating in the last twenty years. These markets include:

Billingsgate: a fish market, now relocated in Dockland

Spitalfields: a fruit, flower and vegetable market, now relocated in Leyton in east London

Smithfield: a meat and poultry market

Leadenhall: once a mixed produce market, now a multi-functional retail arcade

Covent Garden: a fruit and vegetable market, now relocated in the former Nine Elms locomotive depot in south London

Petticoat Lane: a Sunday open-air street market held in Middlesex Street, containing mostly cheap clothing and household goods, and very much part of the tourist circuit; it was originally a Jewish market, close by the former Jewish quarter in Whitechapel.

(Opposite the Custom House turn right and follow sign to St. Dunstan Church, past the church and turn left on Eastcheap)

The City Churches

Evidence of London's commercial wealth, considerable population of old and the former importance of church worship can be seen in the enormous number of churches (of the one, State denomination - the Church of England). The Great Fire of 1666 destroyed 84 out of a total of 109 medieval churches. 54 were rebuilt, of which 51 were the responsibility of Sir Christopher Wren. Over the years more churches were demolished, including eleven as a result of bombing in the Second World War, when all the churches were damaged to some extent, and one very recently from a IRA bomb blast. The blitzed churches left standing, of which St. Dunstan is an example, are practically the only bomb sites remaining in the City; such is the value of land that a memorial to the War in this way cannot be afforded! Today only 47 churches remain, but they include a number of remarkable architectural gems. Many are still places of regular worship, and lunchtime recitals are held for the adjacent workforce. We will pass many of these churches on our walk.

Guilds

The City once contained a large number of different trades, and their merchants and craftsmen organized themselves in the medieval period into livery companies or guilds, something akin to the modern trade union. These guilds include, for example, the fishmongers, grocers, goldsmiths, apothocaries, stationers, brewers, saddlers, haberdashers, drapers, pewterers, and tallow chandlers, trades that are no longer at the forefront of modern society! Each of these guilds has a meeting hall and offices and acts as a private men's club; we will pass a number of these on our walk today. The guilds are also associated with the very archaic form of local government in the City which is more than just ceremonial and has been able to resist the major reforms of the 1880s and 1960s. The City Corporation meets at the Guildhall, and the Lord Mayor is drawn from one of these guilds, as are the aldermen for the various City wards.

(From Eastcheap turn right on Rood Lane, left on Fenchurch Street, right on Lime Street, left into Leadenhall Market and right on Leadenhall Street to the Lloyd's building)

The Financial Quarter

This eastern part of the City, extending from the Bank along Threadneedle Street, Cornhill and Leadenhall Street, and Lombard and Fenchurch Streets, constitutes London's financial quarter (Plate 2.2). However, the growth of these services over the years has presented a serious problem of overconcentration. Land prices have often been astronomic compared to nearby areas; there is serious transport congestion; and expansion has taken place in spite of measures to protect historic buildings and restrict building heights.

In the 1960s and 1970s there was government encouragement for office decentralization to other parts of Britain, including the London suburbs; but the 'big bang' deregulation of financial services in 1986

Plate 2.2   The Financial Quarter of the City of London, looking towards Threadneedle Street with the Bank of England (left), Royal Exchange (right), Stock Exchange (centre) and National Westminster Bank Tower (background) (Photo: H.J. Gayler)

and the greater emphasis of the Thatcher government on the market economy resulted in a massive building boom in the City in the 1980s and an increase in employment. Meanwhile, changing information technology and office needs have rendered a large amount of City offices not suitable for the computer age. The result has been both renewal and decentralization to other sites (for example, Broadgate and Canary Wharf), and a redefinition of the location and extent of London's financial quarter. However, the world recession of the early 1990s has seriously affected the property market, and an over-supply of office space is to be found throughout London.

In the City three very different, but related, activities underpin London's role as a world financial centre: banking, insurance and stockbroking. These activities, in turn, focus on three important institutions: the Bank of England, Lloyd's and the Stock Exchange.

In our walk through the financial quarter we will see that it is still a very male world. The institutions are like clubs, and extensions of male-only public schools, Oxbridge colleges and Guards regiments. Dress codes no longer enforce bowler hats, umbrellas and pin-stripe trousers; but sombre dress, including dark suit and black shoes (with finer variations no doubt denoting one's position), is customary. As women have entered this preserve they have adopted similar formal dress, where the jacket and skirt of the business suit is expected.

Banking

Britain's national bank, the Bank of England, was founded in 1694 and its impressive neo-classical headquarters were constructed at the Bank intersection in the 1720s. It was nationalized by the Labour Government after 1945 and expanded facilities have since been built to the west near St. Paul's Cathedral. Close to the Bank can be found the headquarters of all the major English banks, as well as many foreign, merchant and specialized banks; it must be noted, however, that many banks have chosen not to be in the City, a feature that modern telecommunications can in part overcome. The National Westminster Bank's new headquarters is of particular note: it is the City's tallest building and was a controversial departure in the 1980s from the traditional building height.

Insurance

Britain as a maritime nation became increasingly responsible for insuring the world's ships and their cargoes. One of the leading institutions from the 17th century is Lloyd's in the Leadenhall-Lime

Street area. It started out as a coffee house of that name, and has become a place where groups of underwriters carry on business with insurance agents who in turn represent their clients. An important aspect of business in the City is face-to-face contact; and as Lloyd's has expanded its business, it has been necessary to redevelop its premises. Its most recent expansion in the 1980s, designed by Richard Rogers, is one of London's more controversial buildings and constitutes a major departure in the nature of the streetscape; while it is very functional and contains room for future expansion, it is visually dominant and one of the few buildings in the world (the Pompidou Centre in Paris is another) which is clad in unadorned utilities rather than stone, concrete, brick or glass! Around this area are other insurance firms, as well as numerous shipping offices and other firms dealing with maritime matters.

Stockbroking

The London Stock Exchange, one of the world's leading Exchanges, is to be found east of the Bank of England. Opposite the Bank of England, the Royal Exchange, formerly a museum, is now home to the London International Financial Futures Exchange. Again note the associated functions that cluster nearby.

(From Leadenhall Street turn left on St. Mary Axe, left on Camomile Street, right on Bishopsgate and left into Liverpool Street Station and the Broadgate Centre)

Urban renewal in the City

The opportunities afforded by the wholesale destruction of buildings during the 1940-1941 period, as well as greater government intervention in local planning after 1945, did not result in any extensive change to the layout of the City. Renewal has been considerable but quite piecemeal and largely uninspiring; the legal and spatial ramifications of property hindered largescale developments and changes to the old road pattern. There are four notable exceptions to this piecemeal development where office construction has been on a more massive scale and where the overall plan has turned its back on the medieval street in favour of an off-street, pedestrianized world:

Broadgate

The first large redevelopment scheme we will see on our walk is at Broadgate. It is in fact the largest single development in the City of London since 1666. Begun in 1986, it consists of some 14 buildings, none of them more than 10 storeys high, postmodern in style, containing almost 4 million sq. ft. of state-of-the-art office space. However, its early stages were rocky because it involved the redevelopment of two of London's railway stations. The historic train shed of Liverpool Street was in part demolished and built over (the sop was improved passenger facilities); meanwhile, Broad Street, next door, was made redundant by the diversion of train services to other stations and was then demolished, much to the chagrin of the preservationists. The complex contains offices, shops, artistic and sports facilities, including an outdoor arena/ice rink.

An important aim in the Broadgate development was for the City establishment to undermine the upstart, foreign competition at Canary Wharf where rents were cheaper. By the time of opening, City rents had been lowered and were more than competitive, contributing to the financial failure at Canary Wharf.

(From the Broadgate Centre cross Liverpool Street to Old Broad Street, right on Threadneedle Street past the Stock Exchange, the Royal Exchange, the Bank and the Mansion House, cross into Queen Victoria Street to the Roman Temple of Mithras and turn right on Queen Street; cross Cheapside into King Street, cross Gresham Street to the Guildhall, turn right to Basinghall Street, then left along Basinghall Street to the pedestrian bridge across London Wall to the Barbican)

London Wall

This area between Moorgate and Aldersgate, south of the Barbican, was heavily blitzed in 1940-1941. In 1959-62 part of London Wall was widened to a four-lane arterial road (this was yet another example of piecemeal planning since the road improvements were never completed at either end), and the first high-rise, slab-like, glass and concrete office blocks were built along the route. More recently office blocks have been built across the road space, and walkways above ground level have been built to connect the area to the south with the Barbican.

Barbican

This area north of London Wall was almost totally blitzed in 1940-1941, and the site was a wasteland for almost 30 years. A Comprehensive Development Plan was prepared for the area; and the development, which included changing the route of the Underground lines and most roads, completed in the 1970s, became a showcase for the Corporation of the City of London. The architectural style is Modern but with a heavy emphasis on brutalism - the formidable, concrete, fortress model (shades of Le Corbusier), at a scale that defies close, human contact. Tall high-rise, residential blocks tower over smaller structures, two large ponds, a medieval church and Roman ruins; all of these are contained in a fully pedestrianized area, much of it above ground level and or divorced from contact with the street system (Plate 2.3).

It is very much a mixed-use development. As well as offices, there are apartments; and as a result, for the first time since 1851 the population of the City of London has increased. There are also restaurants and bars, a conference centre, art gallery, concert hall and theatre, a music school, a residence for one of the four major London symphony orchestras, and the London base for the Royal Shakespeare Company. The piazza in the centre and outside seating provide some sense of focus to the development, at least in the warmer weather. The Barbican includes the highly-rated Museum of London which is set up as a walk through the two thousand years of London's history.

(From the Barbican cross London Wall into Aldersgate Street and St. Martin's Le Grand, cross Newgate Street into the Paternoster Square development)

Paternoster Square

This was another 1960's development in an area largely destroyed in 1940-1941. However, it has always been much more controversial. It destroys many of the views of St. Paul's Cathedral to the south and east and does so little to complement Wren's masterpiece. It is reviled as ugly, windswept concrete jungle that has aged poorly, and is an example of what the anti-Modernists, such as Prince Charles, detest about recent architectural design. The offices have been overtaken by technological changes and are now empty. There are plans to tear down the development and to replace it with something more sympathetic to the adjacent St. Paul's Cathedral. However, the

Plate 2.3 The Barbican development in the City of London (Photo: Barbican Centre)

recession in the London property market and the surplus of office space elsewhere have done little to encourage a speedy redevelopment.

(Turn right into St. Paul's Church Yard past the Cathedral and cross to Ludgate Hill)

St. Paul's Cathedral

The western part of the City is dominated by Sir Christopher Wren's Baroque masterpiece, St. Paul's Cathedral, built on the site of the earlier cathedral which was destroyed in the Great Fire of 1666. Miraculously, the cathedral survived the intense fire bombing of the City in 1940-1941 which destroyed much of the property to the north and east. The success of the cathedral as a major tourist feature brings with it the problems of people and traffic congestion, parking problems, especially for tourist buses, and exhaust pollution. For so important a feature, note how building lines on Ludgate Hill (and in particular the Paternoster Square development), as well as the curve of the street, have been allowed to obscure the west front of the cathedral (Plate 2.4).

Plate 2.4 St. Paul's Cathedral from Ludgate Hill, showing the west front partially obscured by the Paternoster Square development (left) and the older shops and offices (right) (Photo: H.J. Gayler)

Ludgate Circus

The bottom of Ludgate Hill marks the boundary between the Roman and medieval City which extends to the west (Figure 2.1). Recently, as a result of the reopening of the north-south British Rail route across central London (the Thameslink service), there has been a modest redevelopment in this area. The rail line has been buried and the rail bridge across Ludgate Hill, together with two older stations, demolished. The City's last sizable bombed-out area, together with the railway lands, are being developed for offices, and a new station (City Thameslink) has been opened to serve the western City area.

(Cross Ludgate Circus and go along Fleet Street, opposite Chancery Lane turn left into the Inner Temple, returning to Fleet Street)

Fleet Street

This area of the City of London was the centre of printing and publishing and a part of the very extensive inner-London industrial belt. All the national newspapers, and their printing presses, were to be found along Fleet Street, between Ludgate Circus and Chancery Lane, or in the streets to the north towards Holborn. Other publishers and the London offices of provincial and foreign publishers were also to be found in the area, as well as a whole range of support services. Until the 1980s the area bustled with activity both day and night; but now it is decaying quietly. In the last ten years all the national newspaper publishers have moved out to better locations, especially in Dockland, in part the result of technological change and the desire for more space, also a demand to break the back of outmoded labour practices. Newspaper printing finished in the Fleet Street area in 1988. The move was perhaps the final stage of decline and relocation that had been going on in the industry for many years before that.

Fleet Street is thus ripe for redevelopment. However, given the surplus of such sites in central London, this one is only going to join the list of future possibilities. Meanwhile, the street and its architecturally-significant buildings, such as the former Daily Telegraph and Daily Express offices, are getting a more despondent air about them as vacancies abound and alternate, and perhaps temporary, uses make little financial commitment to the area.

The Legal Quarter

The very western part of the City of London, and the adjacent area of Holborn (now in the London Borough of Camden), is the legal quarter of London, extending northwards from the Middle and Inner Temple on the Embankment (one of few short stretches of riverside road in London), past the Royal Courts of Justice to Lincoln's Inn and Gray's Inn. These Inns of Court developed after the 13th century and have retained many of the characteristics of the medieval university (which London never received), for example the development around enclosed courts and various social and religious institutions. (In the English context the medieval universities were established at Oxford and Cambridge; London was an early 19th century foundation allowing dissenters (non-Church of England adherents) to attend university.)

The English legal system is two-tier in nature; solicitors represent clients in the lower courts, while barristers, employed by solicitors, represent clients in the higher courts (both civil and criminal). The legal quarter, and its four large Inns of Court, are where barristers are trained (in effect a postgraduate law course) and where many firms have their offices. Nearby are the Royal Courts of Justice, the Old Bailey (the Central Criminal Court), and a host of associated services (legal book shops, wig and robe makers etc.) Other adjacent institutions are closely associated with the law, including the Public Record Office (now relocated to Kew), the Land Registry, the Patent Office and Somerset House (the national registry of all births, marriages and deaths since the 1830s).

Temple Bar

Opposite the Royal Courts of Justice in Fleet Street is Temple Bar, a monument denoting the entrance to the City of London from the City of Westminster, and where traditionally the reigning monarch acknowledges the wealth and power of the City by stopping to gain 'permission' before entering on any ceremonial visit. Because of the inevitable traffic tie-ups, such visits are kept to a minimum and avoided where possible.

(The City of London walk concludes just west of Temple Bar, at the intersection of the Strand and the Aldwych; this is the location where the City of Westminster walk commences)

Further reading

Diamond, D.R. 1991. The City, the 'Big Bang' and Office Development. In Hoggart, K. and Green, D.R. eds. *London: A New Metropolitan Geography*. London: Edward Arnold, 79-94.

Michie, R.C. 1992. *The City of London: continuity and change, 1850-1990*. London: Macmillan.

Porter, R. 1994. *London: A Social History*. Cambridge, MA: Harvard University Press, 11-92.

Pryke, M. 1991. An international city going 'global': spatial change in the City of London. *Environment and Planning D: Society and Space* 9: 197-222.

Punter, J. 1992. Classic Carbuncles and Mean Streets: Contemporary Urban Design and Architecture in Central London. In Thornley, A. ed. *The Crisis of London*. London: Routledge, 69-89.

Chapter 3

City of Westminster

The City of Westminster, to the west of the City of London, has evolved as the administrative centre, including all the major offices in the legislative and executive areas of the British government, as well as the functions of the Head of State and the constitutional monarchy. Associated with these are the diplomatic corps, the offices of countless national and international organizations, military establishments for both ceremonial and guard duties, and some of Britain's most prestigious residential areas. Westminster is also the most important commercial centre for retailing, tourism, entertainment and other non-financial services. The Royal parks meanwhile are the only extensive areas of open space in the central part of London. Westminster lies at the hub of Britain's cultural scene, and is the very pinnacle of its class system.

The second of London's cities dates from the medieval period, in the 10th century, when a monastery and abbey church was built on the island of Thorney in the River Thames, later to become Westminster Abbey. In this period England was developing as a unified nation state, one of the earliest in Europe, and London, and more especially Westminster, became the political capital. It was a role that would continue to grow until the present day. Government focused at first on the Crown, and a succession of royal palaces are to be found in this area. Around the Crown and Government was the Court, a hierarchy of nobility and lay people (Lords and Commons in the Parliamentary sense), who resided close by in their town houses for part of the year. Britain's colonial and industrial might resulted in a political capital that became both highly developed and centralized, attracting a disproportionate share of population and wealth. Central government has continued to take on more powers; in the 1980s, for example, there was considerable legislation which transferred responsibilities from local to central government.

The administrative function dominates the southern part of the present City of Westminster, close to the River Thames and focusing on Whitehall and Parliament Square. Unlike the City of London, this part of Westminster has never become an important hub in the area's transportation network. Only two British Rail terminals come into the fringes of the area, Victoria and Charing Cross; and both serve south and south-east London and the counties of Kent, Surrey and Sussex beyond. Waterloo, serving south-west London and beyond, terminates on the south side of the river, while those parts of London (and England) to the east, west and north have no direct British Rail connections.

Westminster is very reliant on the somewhat slow London Underground for connections with British Rail terminals and access to the suburbs north of the Thames; but even here the situation is far from perfect, and there have been numerous changes (and more proposed) to improve the situation. The focal point of the system has been Charing Cross-Embankment where five Underground lines meet at the two stations (and in close proximity to the British Rail Station). The opening of the Victoria line in the 1960s improved access into the Victoria area from south and north-east London, while the extension of the Jubilee line through Westminster will do the same for north-west, south-east and east London. But parts of the area such as the new office area of Millbank, to the south of Westminster, are on no line at all.

Victoria, which is a large office and tourist-service area of Westminster, has an important rail-local bus terminal interchange. However, London's most important long-distance bus station is inconveniently situated a few blocks to the west, divorced from any rail or Underground connections.

Our walk today will explore various facets of the urban landscape in this administrative centre, as well as the many other features associated with it such as tourism and residential development (Figures 3.1-3.2).

(Commence walk at the eastern end of the Strand near Temple Bar, following the Strand past the Aldwych towards Charing Cross and Trafalgar Square)

The Strand

This Old English word for beach was the name given to the Roman road which led from Ludgate and Fleet Street, following the

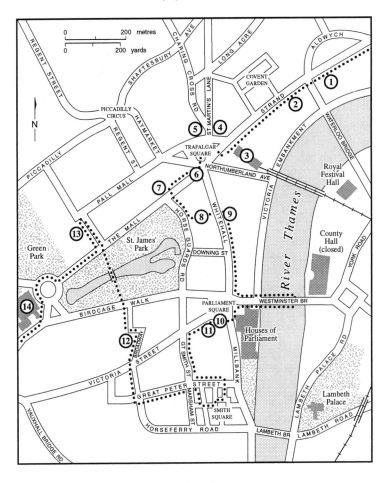

Legend

1	Somerset House	8	Horse Guards Parade
2	Savoy Hotel	9	Banqueting House
3	Charing Cross Station	10	St. Margaret's Church
4	St. Martin's-in-the-Fields	11	Westminster Abbey
5	National Gallery	12	London Regional Transport
6	Admiralty Arch	13	St. James' Palace
7	Carlton House Terrace	14	Buckingham Palace

Figure 3.1 City of Westminster Walk

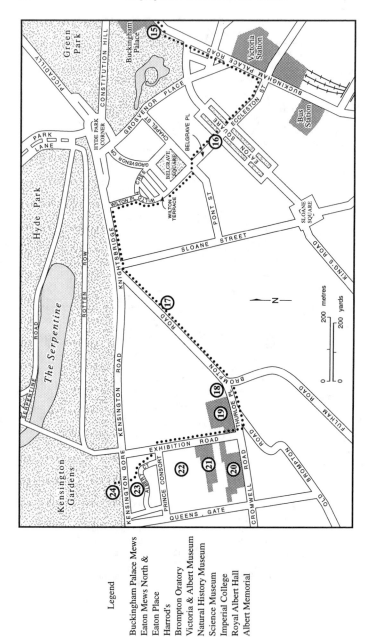

Figure 3.2 City of Westminster Walk (cont.)

Legend

15   Buckingham Palace Mews
16   Eaton Mews North &
17   Eaton Place
18   Harrod's
19   Brompton Oratory
20   Victoria & Albert Museum
21   Natural History Museum
22   Science Museum
23   Imperial College
24   Royal Albert Hall
      Albert Memorial

embankment above the River Thames. Later, it was to connect the medieval City of London with the Court and Parliament at Westminster, and it attracted ribbon development especially homes of the nobility with their gardens extending down to the river. Today the street acts as the southerly extent of the West End, in particular London's entertainment and shopping district, and it contains an incredible mix of functions along its short length.

The Aldwych-Kingsway end represents one of the few attempts at 19th century road improvements across the maze of medieval streets. Together with the approach to Waterloo Bridge, it has become a rather large traffic circle with a number of important institutions in the vicinity, including the British Broadcasting Corporation, the High Commissions for Australia and India, Somerset House, two colleges of the University of London, two churches (forming islands in the Strand), hotels and theatres.

Between the Aldwych and Trafalgar Square can be found a great mix of shops from the seedy to the high-class specialty, theatres, hotels (including the Savoy and the Strand Palace), offices (such as Shell Mex House and various Commonwealth state tourist offices), restaurants (from fast food to Simpson's), a hospital and Charing Cross Station. The Strand is an important through route and the various businesses along its length generate a lot of traffic. It is often highly congested and over the years the Embankment to the south, another 19th century road improvement, has become an important by-pass between the City and Westminster.

Trafalgar Square

With its column commemorating the death of Lord Nelson, Trafalgar Square is the most famous of the London squares (Plate 3.1). It was developed in the late 1820s by John Nash and was part of his scheme (later abandoned) to connect Whitehall and Pall Mall East with Bloomsbury. (Charing Cross Road was constructed on a different alignment in 1887.)

It has since become another of the focal points of the London tourist industry: a place to wile away time, feed the pigeons, meet up with one's friends or visit adjacent sights. It is also an important meeting place for any large march, occasion or demonstration. Whether it be a Ban-the-Bomb rally, welcoming in the New Year or a victory celebration, the crowds make for Trafalgar Square to yell and scream, jump into the fountains and tantalize the police and the pigeons. It has an important place in the British psyche.

Plate 3.1 Trafalgar Square. This view from the forecourt of the National Gallery shows Nelson's column, and in the distance, at the end of Whitehall, Big Ben and the Houses of Parliament (Photo: H.J. Gayler)

Around the square, again a large traffic circle, is an eclectic mix of functions. The most imposing is the neo-Classical National Gallery (1832-38), designed by William Wilkins, who also designed University College, with its controversial extension which was recently opened. To the side of this is the church of St. Martin's-in-the-Fields; the present church was built in the 18th century and is famous now for its concerts and the name it gave to one of England's top chamber orchestras. The High Commissions for Canada and South Africa face the square, the latter until recently with a longstanding anti-government and anti-apartheid vigil outside. The south side of the square has a mix of tourist shops and offices and Admiralty Arch, an office block and gateway to the Mall and Buckingham Palace.

(Cross Trafalgar Square, go under Admiralty Arch into the Mall, turn left on Horse Guards Road, cross Horse Guards Parade, through buildings and turn right into Whitehall)

The Mall

This road leading from Trafalgar Square to Buckingham Palace is the only piece of monumental Baroque town planning in London, and its spacious, tree-lined boulevard bordering St. James' Park is perfect for ceremonial occasions and cavalry regiments going to or from their guard duties. It doubles as a fast by-pass route to the more crowded streets of Westminster and the West End.

On the right of the Mall after Admiralty Arch is Carlton House Terrace, a double row of Regency houses, built by John Nash in the 1820s and now used as government offices. The terrace was constructed on either side of Carlton House which was afterwards demolished to make way for Waterloo Place. The demolition was part of a successful plot by Nash and George IV against Parliament to convince them that Carlton House was beyond repair and to provide funds to convert Buckingham House into a Palace. The terrace was part of a much grander scheme by Nash which will be examined in Chapter 4 below.

Whitehall

This street leading from Trafalgar Square to Westminster is the traditional centre of the executive wing of government, the civil service. Henry VIII was responsible in the early 1500s for extending the seat of government from Westminster to a new Palace of Whitehall (only the Banqueting House remains), and also converting a hospital into the present St. James' Palace. Between Whitehall and St. James' Palace, on land that was confiscated from the Church at the time of the Reformation, the first of the Royal parks was developed, initially as a deer park.

Until the Second World War Whitehall was the centre of all the important government ministries; and a small street off it, Downing Street (now closed to the public for security reasons), contains three houses, the home of the Prime Minister (no.10), the Chancellor of the Exchequer (no.11) and the Government Chief Whip (no.12). The enormous increase in government activities in the last 40 years has resulted in expansions elsewhere in central London, as well as in provincial cities. In fact, Defence, the Foreign Office, the Home Office, the Treasury and the Privy Council are the only major offices that remain. Horse Guards Parade, with its troops in ceremonial dress, and the Cenotaph, the most famous of Britain's war memorials, are important tourist features on this street. Government buildings designed by Sir Gilbert Scott in the Italian style also grace the street.

(Proceed along Whitehall and Parliament Street into Parliament Square. By first turning left into Bridge Street to Westminster Bridge views can be obtained of the River Thames, the City of London and the front of the Houses of Parliament.)

Westminster

This is the centre of the legislative wing of government, although a number of other functions and institutions are found in the immediate area; again it is a major centre of the tourist industry. The various activities that should be noted are as follows:

Houses of Parliament

This impressive neo-Gothic pile designed and built by Sir Charles Barry following the fire in 1834 is not only Britain's largest public building: it easily symbolizes the country's once imperial splendour and centre of its power, especially when seen from the other side of the River Thames (Plate 3.2). This is the original site of the Palace of Westminster, a title that is still used today and is synonymous with the Houses of Parliament. The royal palace dates from the early 11th century, but the only building to survive the fire was Westminster Hall which was constructed in the reign of William II between 1095 and 1100 and considerably altered in the Gothic style 200 years later. The need to incorporate the hall and the adjacent Old Palace Yard resulted in

Plate 3.2 Houses of Parliament from the River Thames, showing Big Ben (right), the Victoria Tower (left) and Westminster Bridge in the middleground. In the right background can be seen the twin towers of Westminster Abbey (Photo: H.J. Gayler)

an irregular shaped building and a less than impressive appearance from the street side.

Big Ben, the clocktower, is the most memorable feature of this building which is made up of the two Houses of Lords and Commons, extensive office accommodation and the residence of the Speaker of the House of Commons. The House of Lords is also the highest court in Britain, and law lords sit to hear appeals. Parliament has long outgrown the building; for example, there is no longer enough office space for the 650 MPs, let alone the over 1,000 members of the Lords. Offices for MPs are to be found in the Norman Shaw building off the Embankment north of Westminster Bridge.

Parliament Square

This is not the place to linger, even if it can be reached! Again, it helps to set off the many important buildings around it, but the place has been reduced to one of London's more congested traffic circles. It is the meeting point of a number of major routes, including Westminster Bridge/the Embankment, Millbank, Victoria Street, Birdcage Walk and Whitehall, and there is little prospect of altering traffic patterns. The continuation of the Embankment under the Houses of Parliament to connect with Millbank would be a logical first step.

Middlesex Guildhall

Facing Parliament across the square is the former administrative headquarters of the old county of Middlesex, constructed in 1913. This part of London is in the historic county of Middlesex (other parts are in Essex, Kent and Surrey). When the County of London was created in 1889 out of the four old counties, Middlesex afterwards set up its headquarters in Westminster outside its new administrative area, testament to the fact that Middlesex really had no focal point other than central London itself. In 1965 the County of London was expanded, becoming Greater London and absorbing the urban parts of Essex, Hertfordshire, Kent and Surrey and all of Middlesex.

St. Margaret's Church

On the third face of the square, in the shadow of Westminster Abbey and very much dwarfed by it, is Westminster's parish church. Its claim to fame is its role as the church for London's elite weddings.

Westminster Abbey

There has been a church on this site since the 10th century, and the present church was commenced soon after 1050, prior to the Norman Conquest. The Abbey is more than a church; there is an extensive site immediately to the south containing related functions. There has been a school here since before 1200; Westminster School is now one of Britain's leading private schools. Since 1540 Westminster has been a cathedral, London's second cathedral after St. Paul's. There are offices of the Dean and Chapter who are responsible for running the cathedral. Church House, the head office of the Church of England, is also here. Indeed, Westminster Abbey is always regarded as the 'Head Church', probably because of its historic associations with royalty; all coronations and most weddings and funerals are held here. Many of the ceremonies are performed by the Archbishop of Canterbury, the ecclesiastical head of the Church of England, who resides not in Canterbury but in Lambeth Palace, across the river from Parliament (see Figure 3.1). Also, the Queen, as Defender of the Faith, is the titular head of the Church, and Britain is one of the few countries with no separation still between Church and State.

Methodist Central Hall

Britain's second religion has its headquarters opposite Westminster Abbey. International classicism in style, dating from just before the First World War, the overall appearance is more auditorium than church.

(From Parliament Square pass the front of Westminster Abbey and turn left on Great Smith Street, left on Great Peter Street, right on Lord North Street into Smith Square, right into Dean Trench Street, right on Tufton Street, left on Great Peter Street, right on Strutton Ground, across Victoria Street to Broadway, right on Queen Anne's Gate and ahead to Birdcage Walk)

Millbank-Westminster

The area to the south of Parliament and the Abbey was opened up in the late 1600s, early 1700s and is home today to a number of important institutions. First, it is still a residential area, focusing on Smith Square and its church, and is particularly favoured by MPs, many of whom have division bells installed in their houses to advise them of a vote in the House of Commons. The headquarters of various political parties are here. Since World War II the area has been invaded by a number of office functions, both public and private. Of note are the three slab-like office buildings on Marsham Street, housing the headquarters of the Department of the Environment, and Vickers Tower on Millbank (the firm are armaments' makers). Numerous hospitals are to be found here, plus the Tate Gallery (London's second art gallery, concentrating on more modern works) and Westminster Cathedral (Britain's chief Roman Catholic Cathedral, built at the end of the 19th century in a somewhat out-of-place Byzantine style).

Victoria Street

The street which connects Parliament Square to Victoria contains a number of office functions, developed principally in the 1960s, and one of the first major attempts to decentralize offices away from the prime rent zones in the City of London and other parts of Westminster. One office function of note is New Scotland Yard. This is the headquarters of the Metropolitan Police, which serves all of Greater London except for the City, and is directly answerable to the Home Secretary in Cabinet. All other police forces are the responsibility of local government and finally to the Solicitor General in Cabinet. There is no national police force in Britain, and in the interests of resisting totalitarianism, it has always been talked out. However, in a number of areas, especially criminal investigations, corporate fraud and national security, Scotland Yard does act as a de facto national force.

Between Victoria Street and St. James' Park is another of Westminster's old neighbourhoods. Once again, favoured as a residential area by parliamentarians, but an interesting mix of functions have since invaded the area, including 55 Broadway (the headquarters of London Regional Transport, designed in the late 1920s by Charles Holden), the Wellington Barracks (the regimental headquarters of five of the Guards regiments, dating from 1833), Caxton Hall (where London's elite goes for out-of-church weddings), and various schools and hotels.

(Cross Birdcage Walk into St. James' Park, cross the Park and the Mall into Marlborough Road to Pall Mall and St. James' Palace)

The Royal Parks: St. James' Park

One of the distinguishing features of London, which enhances its urban quality and sets it apart from other cities, is the extensive area of Crown land, now owned and run by the State, which has remained undeveloped (and will for all time) and is now a series of public parks. These include St. James' Park, which is the most heavily used, Green Park, Hyde Park, which is the largest, and its westerly extension Kensington Gardens and finally, disconnected from the rest, Regent's Park. The grounds of Buckingham Palace are private but originally part of the same land. Beyond central London are extensive areas of Royal and public park, including Greenwich Park/Blackheath, Epping Forest, Hampstead Heath, Richmond Park/Wimbledon Common, Kew Gardens/Old Deer Park, Bushey Park/Hampton Court Park, and Windsor Great Park.

Our walk today will traverse St. James' Park, the only one of the parks to have a strictly formal layout, the responsibility of Charles II (1660-85). In the 1820s John Nash re-landscaped the park and redesigned the canal into a curvy lake as part of his grand scheme.

Facing St. James' Park are a number of residences or former residences, such as Carlton House Terrace, seen earlier. St. James' Palace, begun by Henry VIII in 1531, was the London home of the monarch before Buckingham Palace was built in the 18th century. It is interesting to note that the official title for the monarchy is still the Court of St. James, and it is used when foreign ambassadors are received or British ambassadors are appointed to foreign countries. Clarence House, a wing of the Palace, was built by John Nash in 1825 for the future William IV, and is presently the London residence of Queen Elizabeth, the Queen Mother. Marlborough House was once the home of an earlier Queen Mother, Mary, from 1936-53; along with Lancaster House, it is now used for government offices. Nash had planned residences on the south side of St. James' Park but these were never built.

Pall Mall

Aside from a tendency to be a one-way street/race-track around Piccadilly, this street is part of the old and prestigious residential area of St. James, although today it has principally a mix of office functions. It is noted for being the focus of exclusive and largely male-only clubs, most of which offer dining and bar facilities, libraries and meeting rooms and overnight accommodation. Many of the clubs figure prominently in the political and business worlds, and membership (which is normally by invitation) can enhance both career and social mobility.

(Return via Marlborough Road to the Mall and turn right towards Buckingham Palace)

Buckingham Palace

This is the London residence of the reigning monarch and a part of the seasonal round of homes which include Sandringham in Norfolk (Christmas), Windsor Castle (Easter), Balmoral in the Highlands of Scotland (summer) and a short stay in summer at Hollyroyd House in Edinburgh in order to perform state duties.

The Palace was built for the 1st Duke of Buckingham in 1705 and named Buckingham House. It was considerably enlarged and altered when bought by George III in 1762, and John Nash converted it into a palace for George IV in 1825-30. It has gone through many changes since then, most of which have done little to win over the British public. It has often been argued that the whole spatial concept of the Palace, Queen Victoria Memorial in front, the two park entrances and the Mall is foreign to Britain, and more akin to something that would be found in France (Plate 3.3). (It has to be added that the scale of similar developments in Paris and at Versailles is far grander!)

Plate 3.3 Buckingham Palace from St. James' Park. At the end of the
Mall (right) is the Queen Victoria Memorial in front of the Palace
(Photo: H.J. Gayler)

Buckingham Palace is today a very large office block, as well as
being a residence, since the Queen has constitutional duties to perform
throughout the year which involve a very large staff (e.g. 'approving'
all government legislation, receiving ambassadors, giving out awards
and making appointments, arranging foreign visits etc.).When the
Queen moves home at regular intervals, part of that office block has to
move with her. The Palace is also home to royal children and a large
coterie of staff. It contains priceless treasures, most of which are never
seen by the public, although there is an art gallery for public viewing
of some of the paintings. There are also horses, stables, coach-houses
and garages adjacent in the Royal Mews; and all of this is contained in
an extensive walled garden stretching westwards towards Hyde Park.

The Palace is an important part of the tourist scene with the
Changing of the Guard attracting large crowds whether the Queen is in
residence or not. The area in front of the Palace and the Mall leading to
it perform some of the same roles as Trafalgar Square. Any occasion
involving the Royal Family (for example, a wedding or jubilee of
something) will attract a very large crowd who will expect a balcony

appearance from the monarch and family. The insatiable appetite among the British and foreigner alike for matters regal will insure a crowd outside the Palace at anytime hoping for a glimpse of something of note.

(Follow the left side of the Palace into Buckingham Gate and on to Buckingham Palace Road and Victoria Station)

Victoria

In addition to the office function noted above, the Victoria area has long been an important transportation terminal for London, including a major British Rail station, Underground lines (Circle, District and Victoria), a London Regional Transport bus terminal in front of the station, a long-distance bus station serving all of Britain, a British Airways road terminal from Heathrow Airport (while the rail station contains a terminal from Gatwick Airport). As a result it is a tourist haven and has the largest concentration of tourist services in London. It contains the major information facilities, bureau de change, ticket agencies, and in the immediate vicinity the largest concentration in London of cheap hotels and bed and breakfast establishments. There are also shopping facilities, restaurants, cinemas and theatres.

(From Buckingham Palace Road turn right on Eccleston Street, cross Eaton Square into Belgrave Place, go ahead following the left side of Belgrave Square to Wilton Terrace and Wilton Crescent, turn left on Wilton Place, left on Knightsbridge and left on Brompton Road)

Belgravia

The Royal Parks posed severe restrictions on the westerly expansion of London in the late 18th and early 19th centuries and development had to leap over them. Belgravia represents one of these new western suburbs, although today very much considered a part of central London. It was opened after 1821 by the builder Thomas Cubitt on land owned by Lord Grosvenor. A major attraction was the location close to the new royal residence (Buckingham Palace). Like Mayfair, an older residential area to the north-east, it became an elite area of London, the inner-city address to beat all addresses, and is still favoured by London society. It is also home to many foreign embassies and the headquarters of many organizations and firms. Unlike Mayfair, it has remained largely a residential suburb, or at least retained that appearance; the original houses predominate, masking very often their conversion to non-residential functions.

Belgravia was a speculative development, but it was built to the finest standards for an upper class, with both large single homes and terraces of town houses, fronting wide streets of varying length in grid-iron format with two major squares (Eaton and Belgrave Squares) (Plate 3.4). The landscaping was also superior with pavement, street-lighting, ornamental plants etc. The scale was much grander than Cubitt's other London developments, although the same basic town-house design can still be seen. To the rear of the houses, fronting narrower streets (Mews), can be seen the former stables and living quarters for the grooms. These were converted to garages after 1900; most have now been converted into houses that are very modest compared to those on the major streets but very expensive in terms of the London real-estate market (Plate 3.5). Belgravia and 165 Eaton Place were immortalized in the TV series, 'Upstairs Downstairs,' although the social scene depicted (between 1900 and 1930) has long gone.

Knightsbridge

Together with Brompton Road and Sloane Street, Knightsbridge acts as the local shopping centre for this elite residential area. Shops range from convenience stores found in any neighbourhood to exclusive, high-order stores which attract people from other parts of London, even other parts of England and abroad. Famous business names and fashion designers line the streets. However, the increasing number of tourists has resulted in more down-market chain stores and fast-food establishments opening in the area.

Without doubt the most spectacular store, and the icon for expensive living, is Harrod's; it is England's largest, and by far the classiest, department store, an overly ornate building occupying a complete city block. It is considered a 'must' on the tourist circuit, although the vast crowds do little more than gawk at the merchandise. (We have now left the City of Westminster and are in the Royal Borough of Kensington and Chelsea.)

(Leave Brompton Road at Brompton Oratory, going ahead into Thurloe Place, turn right after the Victoria and Albert Museum on to Exhibition Road, left on Prince Consort Road, right on Albert Court, around the Royal Albert Hall, cross Kensington Gore to the Albert Memorial in Kensington Gardens)

Plate 3.4 Eaton Square in Belgravia (Photo: H.J. Gayler)

Plate 3.5 The Mews behind Eaton Square (Photo: H.J. Gayler)

'Museumland'

This area of South Kensington between Kensington Gardens, Exhibition and Cromwell Roads and Queens Gate contains Britain's foremost concentration of world famous museums, institutes and colleges for the arts and sciences. The idea for such a development came from Albert, the Prince Consort and husband of Queen Victoria, and followed the Great Exhibition of 1851 which was held in the Crystal Palace in nearby Hyde Park. (The Crystal Palace was taken down and reassembled at Sydenham in south London where it served as an amusement park until it was destroyed by fire in the 1930s.) The success of the Exhibition, the millions of visitors that it attracted and the profits that were made led to the Prince Consort's suggestion that a permanent centre should be established for the thousands of exhibits from all parts of the world. He never lived to see his idea come to fruition, but by the end of the century the various institutions had been completed.

Four enormous museums dominate this area - the V & A, (the Victoria and Albert Museum), which contains commercial art and design, china, glass, costumes and furniture; the Natural History Museum; the Science Museum, containing an extensive collection of artifacts of the Industrial Revolution of the 19th century, and best representative of the Great Exhibition itself; and the Geological Museum.

Other institutions include Imperial College (the science and engineering college of the University of London), the Royal Colleges of Music, Art and Organists and the Royal Albert Hall. This ornate, oval concert hall, built as a memorial to the Prince Consort, is today more of a multi-purpose facility, including rock concerts, boxing matches and religious revival meetings. Opposite is the Albert Memorial, the grandest monument probably in all Britain, neo-Gothic in style and the epitome of the High Victorian architecture seen in this area. Close by are the impressive mansions of the upper classes, and some of the first apartments in London, built in the last half of the 19th century. The address, however, has never commanded the same status as Belgravia and Mayfair.

Further reading

Clout, H.D. ed. 1991. *The Times London History Atlas*. London: Times Books.

Harwood, E. and Saint, A. 1991. *London: Exploring England's Heritage*. London: HMSO.

Porter, R. 1994. *London: A Social History.* Cambridge, MA: Harvard University Press.
Summerson, J. 1962. *Georgian London.* rev. edn. Harmondsworth, UK: Penguin Books.

Chapter 4

Regent's Park and the West End

The northern and central parts of the City of Westminster, including the Regent's Park and West End areas, are the focus of this day excursion. This is Post-Restoration London that developed northwards from the Strand and Whitehall after 1660 and was largely completed by the 1850s and the railway era (Figure 2.1). The earlier period was a time of much chaotic development, particularly in areas such as Soho and Covent Garden; the end of the period, in the Regency and the Victorian era, was a time when metropolitan improvements were already being considered. London's boldest and only great plan that ever came to fruition encompasses part of this area.

This is an area of great contrasts: from conspicuous wealth in Mayfair and the Regent's Park terraces to crumbling public housing projects in various parts of the City of Westminster; from spacious layouts in the north to dank, narrow streets in the south, from the crush of middle to low-market shopping on Oxford Street to the quiet opulence of the specialty shops on Bond Street; from the eerie silence of a staid and aging Savile Row to the young, pop-art and noisy Carnaby Street; from the graciousness of Piccadilly to the sleaze of Soho.

The West End is the supreme commercial, retail, tourist and entertainment centre for London and has attracted a large number of office jobs in the non-financial and non-public administration areas. A large daytime working population is augmented by so many tourists during the summer season that the infrastructure of the area (for example, roads, pavements, public transport and restaurant and bar services) is congested to the point of near collapse. Londoners know well enough to stay away if they can! The Regent's Park area to the north contrasts quite markedly; it is more residential, although various office and institutional developments have invaded the area or houses have been converted for these purposes.

Our walk today will examine these different functional areas and urban landscapes, commencing in the north with post-1800 residential development around Regent's Park (much of it resulting from the work of John Nash) and finishing in the West End in Covent Garden, regarded as London's first suburb after 1600 (Figures 4.1-4.2).

(From Euston Square Station turn on North Gower Street, right on Hampstead Road, left on Robert Street, right into Cumberland Market)

The Nash Improvements

An important characteristic in the town planning of London since Roman times has been the concept of incrementalism: London either grew in a very unorganized way (following the major routes out of the City or infilling older developments) or in the organized, often well-conceived, individually-promoted estates such as we have seen in Belgravia and will see later in Victorian and 20th century London. In the last 350 years there have only been three master plans which comprehensively aimed to redevelop London and or set the course for its future growth: Sir Christopher Wren's plans after the Great Fire in 1666, the Nash improvements in the 1820s, and the Abercrombie proposals after 1944. Only the second was to succeed, and as will be seen, it looks very small against the backdrop of a very large city.

The designs of John Nash which we will see today came at the instigation of the Prince Regent, later George IV (1820-30). His interest in urban planning stemmed from a somewhat bitter rivalry with the Emperor Napoleon of France who had grandiose plans for redevelopment in Paris. Also, the Prince Regent, who lived at Carlton House opposite St. James' Park, was interested in building a country villa on Primrose Hill to the north of Regent's Park and wanted a quick and direct route between the two homes. John Nash (1752-1835) had ingratiated himself with the Court and was called upon to execute a grand design.

The design included a number of quite distinct features, including the Hampstead Road area to the east of Regent's Park, Park Village, Regent's Park itself and the adjacent terraces of large homes, Park Crescent, Portland Place and Regent Street, Carlton House Terrace, Pall Mall East-Trafalgar Square and St. James' Park. Some of these have been viewed in Chapter 3 above; the overall scheme will be seen on this walk.

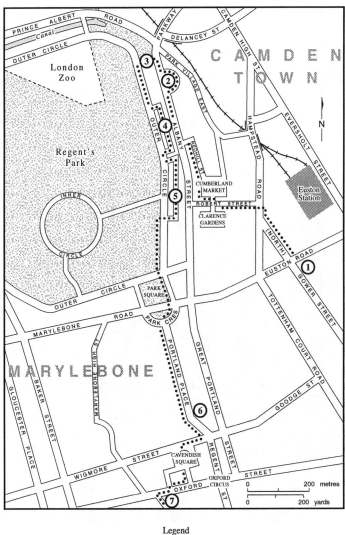

Figure 4.1 Regent's Park and the West End Walk

Legend

1	Euston Square Station	5	Chester Terrace
2	Park Village West	6	All Soul's Church,
3	Gloucester Gate and Terrace		Langham Place
4	Cumberland Terrace	7	New Bond Street

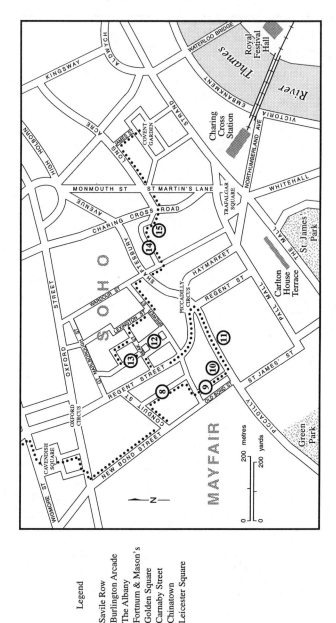

Figure 4.2 Regent's Park and the West End Walk (cont.)

Legend

8	Savile Row
9	Burlington Arcade
10	The Albany
11	Fortnum & Mason's
12	Golden Square
13	Carnaby Street
14	Chinatown
15	Leicester Square

The Hampstead Road area

The first part of the Nash design that we will see is the area between Hampstead Road and Regent's Park around Robert Street. This was laid out by Nash as a working-class complement to the upper-class terraces to the west on the Outer Circle. The model of grid-iron streets and squares (e.g. Robert Street-Clarence Gardens-Cumberland Market) was similar to the upper-class developments to the south. After the Second World War the area was redeveloped as a high-density, high-rise, public-housing scheme with as many as 200 people per residential acre. All of the older housing was torn down, but the original street plan was retained. The quality of much of this recent housing has declined, and renovations and upgrading have taken place.

(From Cumberland Market cross into Red Hill Street, right on Albany Street, right on to Park Village West)

Park Village

Park Village West includes the first examples in 1824 in England of the Italianate villa (a detached house in its own garden) in a suburban setting. (George IV's villa on Primrose Hill was never built.) Nash built similar housing on Park Village East, but some of it was later demolished to make way for railway expansions along the line into Euston Station. After the 1830s the villa would dominate upper-class suburban development, and can be seen first across Regent's Park in St. John's Wood.

(From Park Village West right on Albany Street, left at Gloucester Gate into Regent's Park, turn left on the Outer Circle, left on to Cumberland and Chester Terraces, right at Chester Gate and left on the Outer Circle into Park Square; cross Marylebone Road into Park Crescent)

Regent's Park

Nash was responsible for the enclosing and landscaping of these Crown lands, the northern boundary of which was formed by a branch of the Grand Union Canal; the park was surrounded by an Outer Circle road, and two roads led into an Inner Circle. The idea of having villas throughout the Park was abandoned, although later the London Zoo and the former Bedford College of the University of London were built in the park.

Regent's Park Terraces

Terraces of large homes were built on three sides of the Park facing the Outer Circle. This was an attempt to bring the countryside into the city, and the park-like setting was further enhanced in the case of Cumberland and Chester Terraces by their being set back from the Outer Circle by private gardens. The effect is quite dramatic; the massive neo-Classical developments, some with entrance archways, carved pediments and columns, convey a magnificence and pretentiousness not seen in London's other upper-class residential developments (Plate 4.1). Within the terraces are narrow town houses, most today divided into expensive, highly sought after apartments. The properties are uniform in their appearance as a result of being well maintained by the Crown Estate.

Park Square and Park Crescent

Park Crescent, dating from 1812, was the southern half of a projected circus, and modelled on the Royal Crescent in Bath. Later it was extended across the New Road (now Marylebone Road) as Park Square, and both features acted as the grand entrances to Regent's Park and Portland Place.

(Follow Portland Place into Langham Place, right on Cavendish Place, cross Cavendish Square and follow Holles Street, turning right into Oxford Street)

Portland Place/Regent Street

Portland Place was laid out after 1774 and incorporated into Nash's plans. It leads to Regent Street, but in almost typical English fashion Nash very deliberately did not make the two roads line up. The kink in the road at Langham Place allowed for All Soul's Church to be built. South of Oxford Street the new road cut across the narrow streets of the early 18th century developments in Mayfair and Soho. Later, Regent Street became a social divide between these two very different areas.

The southern end of Regent Street, seen later on this walk, curves eastwards towards Piccadilly Circus in order to line up with Lower Regent Street and Carlton House at the end of the street. The curve fortuitously focuses attention on the buildings, rather than the endless vista of the street. Very little of what Nash built can be seen today because of the redevelopment, some would say architectural vandalism, of the 1920s (Plate 4.2).

Plate 4.1 Cumberland Terrace (Photo: H.J. Gayler)

Plate 4.2 Regent Street near Piccadilly Circus (Photo: H.J. Gayler)

The Marylebone area

Extending north of Oxford Street is Marylebone, once a medieval village outside London (note the old High Street that cuts across the later grid-iron plan in Figure 4.1). In 1889 it became a Metropolitan Borough within the new County of London, and in 1965 it was absorbed into the larger City of Westminster. In the 1700s this area was laid out as a series of estates on land that belonged to England's aristocracy (e.g. the Cavendish-Harley Estate). Plans were similar to other 18th century developments, such as grid-iron streets of various lengths, incorporating different size blocks and open squares of greenery, although there are fewer squares here and more through streets. It was laid out as a residential area, and this is still much in evidence today. However, over the years non-residential functions have invaded the area.

To the west of Tottenham Court Road and east of Portland Place/Regent Street was another of London's inner-city industrial areas, developing after the 1850s; the manufacture of clothing, especially tailoring and women's outerwear, was small-scale and carried out in converted residences. Associated with this was warehousing, showrooms, designers, offices and the export trade. Increasingly, there was direct selling to the retailer; and as this area was close to the emerging retail strip of Oxford Street, there was an in-migration of firms from other parts of London, in particular from the clothing areas of the inner East-End near Bishopsgate and Aldgate. The emergence of the mass-produced factory clothing article for the middle and low-market inevitably led to a decline in much of this inner-area clothing trade.

The area to the west of Portland Place/Regent Street had fewer clothing firms. It was more residential and attracted a number of institutions and professional services. Harley Street, one block to the west of Portland Place, has long been noted as a location for medical specialists, mostly in private practice.

Oxford Street/Regent Street

These are probably world-known as major shopping streets, and are what readily comes to mind when one talks of the West End. Retailing of the middle to low-market type found here contributes to Oxford Circus being the busiest station (in terms of passengers) on the London Underground system. It acts as the highest-order shopping centre in the national as well as the Metropolitan context; indeed, it has inhibited the growth of certain types of retailing in different parts of London. It has long suffered from very serious congestion, both on the road and the pavement, and private car traffic is banned for most of its length between 07.00 and 19.00, Monday-Saturday.

Oxford Street emerged as a prime retail area after 1900 when the Central Line of the Underground was opened under the street. In 1907 the Bakerloo Line was built under Regent Street, intersecting at Oxford Circus. By the Second World War the area between Tottenham Court Road and Marble Arch, and Oxford Circus to Piccadilly Circus was a strip of major department stores (the most notable buildings being Selfridge's and Liberty's, the latter designed in the 1920s as a neo-Tudor building with a neo-Classical facade!), the chief stores of the clothing and footwear chains (for example, Marks & Spencer's), specialty stores for clothing, china and glass, jewelry, recorded music, and furniture, and associated services such as banking (Plate 4.3).

The character of these two streets has changed in the last 40 years, reflecting 1) the decentralization of population in Greater London and a greater percentage therefore residing further away from the West End; 2) the greater strength of high-order suburban centres, including the opening of Britain's first regional shopping centre at Brent Cross in 1976; and 3) the increase in the tourist population of London, and the emphasis on quality goods for direct export, clothing, footwear and other easily exportable items, tourist bric-a-brac (tee-shirts, Union-Jack embossed household articles etc.), and travel services (travel agents, airline offices, bureau de change etc.). Nearly 20 percent of total sales on Oxford Street are reckoned to come from foreign tourists; and the summer tourist mecca is no doubt a deterrent to Londoners shopping in the area.

(From Oxford Street turn left on New Bond Street, left on Conduit Street, right on Savile Row, right on Burlington Gardens, left through the Burlington Arcade (if closed turn left at Old Bond Street), left at Piccadilly)

Mayfair

This area, bounded by Oxford Street, Regent Street, Park Lane/Hyde Park and Piccadilly, is synonymous with wealth, power and influence, where the elite live, work, shop, play and partake of culture and where the elite of other countries come to do the same things. This old residential area was opened up in the late 1600s and early 1700s with the traditional pattern of squares and connecting streets, but over the years it has been heavily invaded by various non-residential land uses.

Numerous office functions are found, especially those associated with the information gathering and processing sector. Mayfair, like

Plate 4.3   Oxford Street with the neo-Classical facade of Selfridge's
department store (right) (Photo: H.J. Gayler)

other parts of the West End, became an office area after 1900.
Structural changes in London's economy and its role as the capital of
the leading world power promoted this development, including the
tremendous increase in professional organizations and institutes in law,
medicine, business, education and journalism.   National firms in
industry and services looked to London for head offices.   There was
little need to be in the City of London close to financial
establishments; indeed, it was better to be closer to government in
Westminster or foreign embassies.   Many of these firms preferred to
occupy residential properties; certainly rents here were cheaper than in
the City.   Together with Belgravia, Mayfair is a major diplomatic area.
The U.S. Embassy is in effect a very large modern office block at the
west side of Grosvenor Square.   Cultural aspects are important here,
including the Royal Academy of Arts on Piccadilly (which features
major exhibitions) and numerous galleries.   There is a concentration of
London's oldest and most expensive hotels (Claridge's, Dorchester, the
Hilton (London's first high-rise hotel), and across the street in
Piccadilly, the Ritz).

As in the area north of Oxford Street, the eastern half of Mayfair was
an important clothing manufacturing area. This was associated with the

Oxford Street retail trade, but more particularly the specialized tailoring to be found in the Savile Row area. So much of this continues to be hand-made to order, rather than factory-made, and the small workshop industry continues to flourish in the vicinity.

So many of the industries and services are present because of the specialized retail trade. Bond Street is one of the world's most exclusive shopping streets, including famous names in the world of clothing, jewelry, art dealers etc. It is a world away from Oxford Street around the corner. Savile Row is world-famous for its specialized tailoring establishments; some of its retailers now spend part of the year on an international circuit (setting up shop for a few days in exclusive foreign hotels). Hardy Amies, the Queen's couturier, is perhaps the best known name to be found here.

A charming aspect to retailing in British cities, and a precursor to the modern shopping mall, is the arcade. Few were built, and fewer now remain. One of the best examples, and the most exclusive, is the Burlington Arcade, built in 1819. Nearby is the Albany, a court of exclusive apartments for men, dating from the early 19th century. The poet Byron lived here, and the former Prime Minister Edward Heath has a London home here.

(Along Piccadilly to Piccadilly Circus)

Piccadilly

This famous street of shops, offices, hotels and institutions leads from Hyde Park Corner to Piccadilly Circus. Fortnum and Mason's is probably the world's most renown grocery shop (founded in 1701), and is by appointment to royalty and long patronized by the upper class. One can be served by men in morning dress and services include, for example, picnic hampers for the Races or Glynebourne.

Piccadilly Circus

The walk down Piccadilly will catapult one from the exclusiveness of Mayfair back into the hype, crowds and confusion of London's West End at Piccadilly Circus. It is a different West End though from that seen earlier on Oxford Street. Piccadilly Circus is the beginning of the entertainment district of theatres and cinemas, which fans out east and south along Haymarket, Shaftesbury Avenue and Coventry Street/Leicester Square towards Charing Cross Road and the Strand. Associated with this is an extensive range of restaurants, bars and nightclubs to fit almost any taste.

The area is probably the chief focal point for tourists, and hordes of young people especially lay about the statue of Eros or in Leicester Square nearby. The major entertainment for many is people-watching and interacting with others, both of which come very cheaply! It is one of the few large areas of London that does not come to a grinding halt when the shops and offices close at 6 p.m.(late-night shopping is a relatively recent event in Britain). However, London's nightlife does not extend much past 11 p.m. when the bars close! The illuminated advertisements in and around the Circus are an important attraction in their own right, and the area should be experienced by night as well as by day.

(Turn left on Regent Street, right on Beak Street, left on Carnaby Street, right on Ganton Street, right on Marshall Street, left on Beak Street, right on Lexington Street, left on Brewer Street, right on Wardour Street, cross Shaftesbury Avenue and left on Gerrard Street, right on Gerrard Place, left on Lisle Street to Charing Cross Road)

Soho

The area bordered by Oxford Street, Regent Street, Coventry Street/Leicester Square and Charing Cross Road is known as Soho. It was one of the West End's first large residential areas, being built principally in the period 1680-1700, and prior to the gracious estate building of the 18th century. The area is almost uniformly grid-iron in design, but unlike later developments the streets are much narrower and there are only two squares (Soho and Golden Squares), neither of which provide a focal point. 19th century road improvements barely touched the area; both Shaftesbury Avenue and Charing Cross Road were laid out in the 1870s across this maze of streets but the aim was to improve access to the West End.

For years Soho has been synonymous with poverty, sleaze, crime and the underworld, and the various trades and peoples attracted to the area either engaged in these things, or were perceived by the British to do so! From the 1660s onward Soho became a gateway for immigrants from other parts of Europe, for example the French Protestants (or Huguenots), many of whom were silk weavers. As Britain's colonial power increased, immigrants came from other parts of the world; a small Chinatown, for example, can be seen in the southern part of the area, focusing on Gerrard Street. Many of the immigrant groups became associated with the food, restaurant and entertainment business; in an era of the bland British meal, Soho was the place to go for good foreign food. It still is! Over the years, and especially since foreign immigration was limited after 1911, this gateway role has decreased. Also other parts of London assumed the role as well.

Soho was also part of this inner-London industrial belt, again focusing on the clothing trade and its nearby retail establishments. Like all these industrial areas the 1960-90 deindustrialization of London has been devastating. Close by the theatres and cinemas, Soho has many associated industries and services. In the 1920s and 1930s the film industry was centered on Wardour Street; the television industry grew up around the corner on Frith Street, where John Baird, the father of British television, lived. Numerous other theatrical services can be seen in the area.

Soho has long been recognized as London's Red Light district. In a nation imbued with puritanism and the Protestant work ethic, this aspect of the capital's life, which no legislation can effectively cope with, has always been quietly pushed under the carpet. Soho has long been the home of the strip-joint, the after-hours bar, the nightclub and the brothel with its pimps and prostitutes; and its juxtaposition with tourism make for a constant flow of customers.

Carnaby Street in Soho is part of London folklore, epitomizing the Swingin' Sixties. At this time laws and codes of conduct became more liberal and this street attracted a large number of clothes boutiques, particularly 'mod' clothes for teenagers, which were considered 'way-out' by older generations. Since that time fashions for clothes have expanded and diffused: now the Kings Road in Chelsea might be considered the place for the young in the 'Swingin' Nineties' to go. However, thousands flock each day to Carnaby Street, but it is as much tourist junk perhaps as teen fashion, although there have been attempts of late to reintroduce a fashion trade.

Charing Cross Road

This major road into the West End again exemplifies the spatial concentration of a specialized service into a small area. It is by no means one of London's major shopping streets, but for a long time it has been the focus of the new and used book trade. The world's largest family-run store, Foyle's, is located here. The street's other role is its theatres, cinemas, restaurants and associated services

(Cross Charing Cross Road and St. Martin's Lane into Long Acre, right on James Street into Covent Garden)

Covent Garden

The area between the Strand and Long Acre constitutes London's first suburb and the beginnings of the West End. In 1631, John Russell, the 4th Earl of Bedford, whose London house was close by, commissioned Inigo Jones, who was Surveyor to the King, to lay out a square and surrounding streets on the former monastic gardens at Covent Garden. The first church to be built since the Reformation, St. Paul's, was opened opposite the square. Jones had been to Italy and imported many ideas, including Palladian designs, from there. The square was in fact a piazza; the rows of houses around the square were without ornament, employed stucco for the first time and were entered by a columned arcade across the front. The houses had narrow gardens with stables at the end.

It was the start of a 200 year phase in London's development of formal, often superior quality, town planning that was based on the square of varying dimensions with connecting streets, the two to five-storey housing being built in rows or terraces. Jones' piazza was not to survive; the square was filled with a fruit and vegetable market in the 1670s and the arcades are long gone, although reproductions were constructed in the 1880s. What had been an elegant residential area became a slum, and aristocratic families left for Mayfair and Bloomsbury. In the 1850s the present market buildings were erected in the square. Adjacent to the market was built the Covent Garden Theatre, later rebuilt as an Opera House. For years the sight of opera-goers in evening dress picking their way through the remains of the day's market always seemed incongruous; and the location and magnificence of the Opera House itself, crammed into a maze of narrow back streets, is a far cry, one might add, from the locational splendour of many of its European counterparts.

The area began to change considerably after 1974 when the fruit and vegetable market moved to the site of the former Nine Elms locomotive depot at Vauxhall. Plans to demolish the market buildings were abandoned after public protests, and the now-defunct Greater London Council developed the area with an exciting mix of cafes, boutiques, craft shops and market stalls. The London Transport Museum was moved here, and artists regularly perform for the largely tourist crowd. Covent Garden has become an extension of the West End tourist scene; and this type of 1980's retail and tourist development is being replicated at other historic sites in Britain.

Further reading

Hall, P. 1962. *The Industries of London since 1861.* London: Hutchinson.

Harwood, E. and Saint, A. 1991. *London: Exploring England's Heritage.* London: HMSO.

Lawrence, H.W. 1993. The Greening of the Squares of London: Transformation of Urban Landscapes and Ideals. *Annals of the Association of American Geographers* 83: 90-118.

Morgan, B.S. 1991. The Emerging Retail Structure. In Hoggart, K. and Green, D.R. eds. *London: A New Metropolitan Geography.* London: Edward Arnold, 123-140.

Porter, R. 1994. *London: A Social History.* Cambridge, MA: Harvard University Press, 93-130.

Summerson, J. 1962. *Georgian London.* rev. edn. Harmondsworth, UK: Penguin Books.

Chapter 5

Bloomsbury and St. Pancras

Bloomsbury and St. Pancras are two inner-London suburbs and constitute the southern part of the London Borough of Camden. The area developed northwards from the Strand and Fleet Street between the 17th and early 19th centuries, largely as a middle-class residential estate on the lands of the Dukes of Bedford. Since the 19th century a wide range of non-residential functions have been attracted to the area, reflecting London's growth as a major world city, good transportation facilities in the area, and the proximity of Bloomsbury and St. Pancras to the centre of the city (Figure 2.1).

Socially and economically, the area is divided by the main north-south thoroughfare (Upper Woburn Place, Woburn Place and Southampton Row). To the west, the residential function has been considerably overtaken by other functions. The headquarters of the University of London at Senate House are here, and close by are various institutes, schools and colleges of the University, as well as numerous other educational functions. One of these is the teaching hospital at University College and the large number of specialist hospitals in the area. There is also an important library and museum role: the British Museum and British Library are located here, although the about-to-be-completed British Library is to be hived off to new quarters further north.

This part of Bloomsbury is also one of inner London's tourist accommodation areas and contains many office functions, including the headquarters of important institutions such as the British Medical Association. The University presence has resulted in residential change. While some middle-class housing is still to be found, either in older houses or more recent blocks of flats, this area is viewed as the student

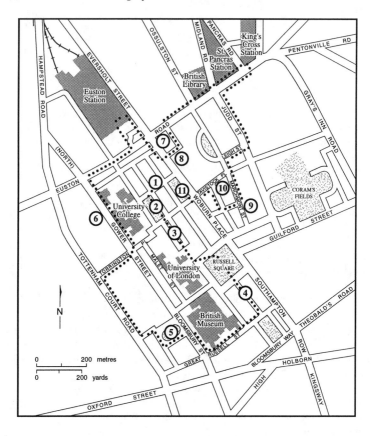

Legend

1	Passfield Hall, Endsleigh Place	7	St. Pancras Parish Church
2	Gordon Square	8	Woburn Walk
3	Woburn Square	9	Brunswick Shopping Centre
4	Bedford Place	10	Peabody Estate
5	Bedford Square	11	Tavistock Square
6	University College Hospital		

Figure 5.1 Bloomsbury and St. Pancras Walk

quarter of London, with many residences near the various colleges of the University. Early in the 20th century the presence of a university resulted in the area becoming the home of the Bloomsbury Group, a loosely-formed group of intellectuals, writers, university teachers and social reformers.

The eastern part of the area contrasts quite markedly with the west. Indeed, many would not regard it still as Bloomsbury, but rather St. Pancras in the north and Holborn in the south, reflecting the centres of the pre-1965 metropolitan boroughs of the same names. It will be seen that the area is more residential and far less grand than the area to the west, but there is more of neighbourhood feeling, focusing on a local shopping street. Again, a number of specialist functions, such as hospitals, tourism and transportation, are to be found.

Bloomsbury has a well defined northern boundary, coinciding with the Euston Road and approximating the extent of London in 1820 (Figure 2.1). The Euston Road marks the southern limit of main-line rail access into this part of London, and to the north of here are to be found major rail termini, servicing facilities and working-class residential areas.

Our walk through Bloomsbury and St. Pancras will enable us to appreciate the urban landscape of inner London (Figure 5.1). We will examine the housing, land tenure and urban form of Georgian and Regency development, the major functions and institutions that have developed since the 1800s and some of the major planning issues that affect the area.

(The excursion begins in Endsleigh Place, situated between the northern ends of Tavistock and Gordon Squares, one block to the west of Upper Woburn Place or two blocks south of Euston Station)

Georgian and Regency housing

The north side of Endsleigh Place gives us an opportunity to examine the housing that is typical of this period (Plate 5.1). It was once a terrace of seven individual homes (some of the former front doors are still visible), but more recently converted into one of the residences of the London School of Economics (LSE), a college of the University of London.

This area is part of an arc to the north and west of the centre where London's first middle and upper-class suburbs developed between the mid-17th and early 19th centuries (Figure 2.1). Georgian refers to the reigns of George I, II, III and IV after 1714; the Regency is the period

Plate 5.1 Passfield Hall, Endsleigh Place. Note how much grander the middle and two end homes were, compared to the four in between, resulting from more decorative windows or the use of the Classical pillar design on the building facade. Not seen here are the more splendid side entrances to the end homes (Photo: H.J. Gayler)

1811-30 at the end of the reign of George III (1760-1820), when his son, the Prince of Wales, took over various administrative roles, and later in his own right as George IV (1820-1830).

Georgian and Regency housing, often called Palladian, is distinguished by a very simplistic, clean, rectangular form with subtle trappings of neo-Classical design (pillars, cornices, capitals etc.), made out of local building materials. Modest working class housing of the two-storey variety was never very extensive and is now rare anywhere in London; it has been subject over the years to massive slum clearance and redevelopment. However, long terraces of middle-class houses, of which Bloomsbury is typical, can still be seen, and great emphasis is now placed on preserving this 18th and early 19th-century urban landscape.

Middle-class housing was usually four or five storey, and the latter describes the housing in Endsleigh Place. Basement level was kitchen, scullery, servants' day quarters, store rooms etc. with direct access by stairs to the street for the servants; the iron railing was not only for safety but maximizing natural light for the basement rooms. A front door for the family of the house led to the ground floor. This would contain the normal daytime living rooms and included the morning room, study, dining room, hall, and cloakroom. A grand front staircase would lead to the first floor where the family would entertain (the drawing room), hence the grander appearance of this floor still seen in its taller windows and balconies. The second floor would consist of bedrooms, sitting rooms, dressing rooms and bathrooms for the family. The third floor would contain the servants' bedrooms, the nursery and accommodation for nursery maids; this floor may be called the attic and is separated architecturally from the floor below (horizontal stonework, smaller windows etc.). The servants' living and sleeping quarters would be connected in the larger houses by a backstairs. There was no garden for such a 'town' house; the yard or 'area' would be for servant use and storage. (In the much grander housing seen in Chapter 3 above in Belgravia the yard would lead to servant housing and stables for horses and carriages facing a narrow street, or Mews, behind.)

Thomas Cubitt was mostly responsible for developing this part of Bloomsbury in the 1820s. Although the house building was highly speculative, the results were substantial and praiseworthy. Summerson (1962, 192) draws our attention to the ornamentation and sense of continuity presented by the ends of the terraces facing Tavistock and Gordon Squares which in turn form the south side of Endsleigh Place.

(Proceed west from Endsleigh Place and turn left into Gordon Square, go along or through the square, cross Tavistock Place

into Woburn Square, and at the end take the pedestrian route into Russell Square)

Georgian and Regency Squares

An important part of the residential fabric was the use of a city block for a small park; it constituted a formal garden with paths, a place to relax and for children with their nannies to play; there would be no vigorous sports allowed. All squares were originally private, fenced and gated, with nearby residents having a key for access. Many are now public, but the fences and gates often remain to enhance security after dusk and prevent people sleeping outdoors.

Subtle differences can be seen in the housing in the area, denoting perhaps different builders, dates and housing costs. Invariably, much more imposing housing would face the square compared to adjacent streets. No two squares are alike; the original plans resulted in different designs and sizes (Russell Square is the largest of the Bloomsbury squares, whilst Woburn Square hardly befits the title), and subsequent development has greatly altered their character in many instances. Residents' cars fill the perimeter, many individual homes have become offices or student residences, and so much of the Georgian and Regency housing has been torn down for newer and more functional buildings.

Land Tenure

The development of streets, squares and buildings in this part of London constitutes what many have called the Golden Age of town planning in the capital. Much of the development of inner London took place on the estates of England's landed gentry, the Church of England and other institutions (for example, the colleges of the Universities of Oxford and Cambridge). These owners did not develop piecemeal but had architects, surveyors and builders lay out extensive estates at one point in time and build to that plan in subsequent years.

These estates are still largely owned by the same people and institutions, and the land is leased to the owners of property situated on it. Leases are long term, 99 years is popular; failure to renew a lease can mean the loss of one's investment in a property, and the length of lease remaining has a major effect on property prices. Prior to the days of land-use planning these estate companies had important controls over how an area developed. A large part of Bloomsbury was, or is, owned by the Duke of Bedford, hence the many street names, building names etc. relating to that family - Bedford, the Dukedom itself; Woburn, after the country estate, Woburn Abbey, in Bedfordshire; Russell, the family name; and Tavistock, after the Marquis of Tavistock, the Duke's eldest son.

The Russell Square area

This large square lies at the very heart of Bloomsbury and was no doubt once the showpiece of the estate. Since the early 19th century various non-residential functions have moved into the area, and most of the original housing has gone, although great pains have been taken recently to build a neo-Georgian property on the west side of the square! To the east of the square, and extending along Woburn Place and Southampton Row, is one of London's major hotel districts, serving the middle-market foreign tourist on a package holiday, as well as being close to three major railway stations and tourists from other parts of Britain. The hotels vary from the opulence of late 19th-century neo-Gothic to 1960's Modern of no particular architectural significance. Many tourist-oriented services are to be found in the area; and to the north-east in the Kings Cross-St. Pancras area is an extensive district of small bed and breakfast type hotels. Sadly, a major north-south arterial route passing through the area has resulted in the square becoming a vast traffic roundabout, and tour buses outside the hotels cause congestion.

University of London

Abutting the north and west sides of Russell Square is the University of London, the largest single non-residential land-use in Bloomsbury. Its structure is a central administration and library, run from the imposing, white stone, 1930s-era Senate House (which backs on to the square), and a lower-tier system of colleges, institutes and schools where the actual teaching, research etc. is carried out. Some of these are close by and can be seen on the walk, e.g. University College, and its teaching hospital, the Institute of Education, the School of Oriental and African Studies, the School of Slavonic and East European Studies, and numerous smaller schools and institutes.

There are a host of associated services, for example, the Student Union building, Dillon's Bookshop, art galleries and theatres, student residences (both University and College), and other teaching and research hospitals and medical facilities. (The following specialist hospitals can be seen in the area: Foot, Dental, Ear, Eye, Nervous Diseases, Homeopathic, and Children.)

The growth of higher education has resulted in the University of London being an avid developer, and most of the new buildings have been at the expense of the Georgian and Regency terraces. Conflict eventually arose with those organizations which wished to preserve the heritage of the area; the emphasis now is on the University adapting the existing housing stock to its use since demolition and redevelopment are often not possible. In some cases this means development outside the area, which is not without merit. Six of the largest colleges are not to be found in this area; together with an array of other institutes, they can be found up to 50 miles away.

(Leave Russell Square on the south side, cross into Bedford Place, turn right on to Great Russell Street to the front of the British Museum)

Bedford Place

This relatively quiet street, complete with original and recently refurbished properties, is perhaps the best example in the area for gaining an impression of the Bedford Estate street as intended.

British Museum

Extending north from Great Russell Street and abutting the southwest corner of Russell Square is Britain's largest museum and library, an impressive, neo-Classical, 19th-century building fronting a decidedly unimpressive street. The museum specializes in early and non-British artifacts (and contains, for example, the controversial Elgin Marbles). The library with its famous circular reading room for researchers is one of Britain's statutory libraries which since 1911 has to receive every book published in the country. There is also an large map collection. For many years the museum and library have run out of space and various collections are housed elsewhere (for example, newspapers in a warehouse at Colindale in north London; some things have even gone to Yorkshire!). There is no room for expansion on the old site, short of tearing the building down. After much argument and delay, a new British Library is being constructed on the former Somers Town Goods Station on Euston Road to the west of St. Pancras Station. Unfortunately, the building is already full before it opens!

(From the British Museum turn right along Bloomsbury Street, left into Bedford Square, cross the square and exit on Bayley Street and turn right on to Tottenham Court Road)

Bedford Square

This is regarded by many as the best example of a Georgian square in London. It is complete and the original design of the 1770s is little changed. The square is basically symmetrical, and along each face the centre house is larger and has a more imposing Classical facade than its neighbours. The square contains an oval-shaped park. A factor in preserving this square (and Georgian architecture in general) is probably the Architectural Association which for nearly 80 years has had its headquarters here. Unfortunately like so many London squares, the initial calm has long since been broken by a major traffic artery along its eastern face.

Tottenham Court Road

This major arterial route forms an approximate western boundary to Bloomsbury. It acts as an extension (albeit a contracting one) to the West End shopping area along Oxford Street, and developed in the early part of the 20th century following the opening of the Central and Northern lines of the London Underground (the Northern Line follows Tottenham Court Road). Its major retail role was the furniture district with a number of very large stores for that purpose. Of late, however, this role has suburbanized, and the University, medical and specialty retail functions have taken over.

(Turn right along Torrington Place, left on to Gower Street and turn right on to Euston Road)

University College

This is the oldest, as well as one of the largest, of the colleges of the University of London. Its original building, dating from the 1830s, which is set back from Gower Street across a courtyard, is from the same mould (by the same architect) as the National Gallery in Trafalgar Square. In the more recent past, as higher education has expanded, the college has taken over more and more of the area, but a congested and restricted inner-city site has not allowed for particularly inspiring developments. The College is constrained by the original Bedford Estate plan, and since the 1960s there has been much greater preservation of the Georgian and Regency developments.

A faculty of medicine and a teaching hospital have been important functions of University College since the 1830s, but in the 1890s it was realized that a completely updated and larger hospital was needed.

The result was the University College Hospital seen today on Gower Street. While the airiness of the design was championed, its style, height and overwhelming red-brick appearance were considered intrusive. Moreover, the design of four wings radiating out from a central service area was based on outmoded ideas of disease prevention, and was an inefficient use of a small site.

Population decline in inner-London has resulted in there being less need for hospital bed space, especially for routine procedures, and bringing in patients from elsewhere is often costly and inconvenient. The outcome has been the on-going program of rationalization and closure of facilities, which in turn has led to threats and uncertainties to world-renown medical programs and opposition from both the medical establishment and community groups in some of London's poorest areas.

Euston Road

In the 1750s, the New Road, a by-pass from Paddington to Islington, was opened to the north of the then built-up area, and later extended as City Road towards Shoreditch. Today, part of this route, Euston Road, forms the northern boundary to Bloomsbury; and its role as a by-pass (albeit very much inner-city) still applies, connecting the M40/M4 highways to the west of London with routes such as the A1, M11 and A12 to the north and east. Some intersections have flyovers to improve traffic flow, and improved accessibility has resulted in a variety of office functions being attracted to the area after the 1960s.

Euston Road and its westerly extension, Marylebone Road, were once the main streets of their respective communities of St. Pancras and Marylebone. The town hall of the former Metropolitan Borough of St. Pancras (now the London Borough of Camden) is located opposite St. Pancras Station; while St. Pancras Parish Church, dating from the first housing developments in the 1815-1825 period, is situated at the eastern end of Euston Square. The sheer volume of traffic today, together with its noise and pollution, result in Euston Road providing little in the way of a community focus and essentially fracturing the neighbourhoods to the north and south.

(Right along Euston Road crossing to the north side at Gordon and Melton Streets, follow the signs across Euston Square into Euston Station)

London's Railways

Euston Road was the approximate northern limit of urban development at the time the first railways reached London in the 1830s (the London and Birmingham Railway, later the London & North Western Railway, into Euston). Euston, St. Pancras and Kings Cross Stations typify the ring of London termini beyond which the railways were not allowed to penetrate any further into the centre. As it was, hundreds of homes (usually poor ones) had to be demolished to let the railways come this far in. In the central area there is only one through north-south British Rail line; this had been closed for years until it was re-opened by British Rail in 1988. Two east-west through lines exist; but they are now part of the London Underground network, and there have been no long-distance through trains since the Second World War.

All the individual railway companies were involved in goods (freight) transportation which was designed to serve the needs of London's people and businesses. Two companies, the Midland Railway and the Great Northern Railway, developed facilities at Somers Town and Kings Cross respectively. Since the vast majority of goods transportation is now by road, these facilities have been closed. Somers Town is the site of the new British Library, while the Kings Cross facility awaits redevelopment. Likewise, each of the railway companies needed its own locomotive and carriage servicing facilities, and these are still to be found to the north of this area. The railway companies serving this part of London were once major employers, and extensive areas of the working-class housing were built to the north of the Euston Road. More recent slum clearance programs have resulted in various public housing developments in the area.

Euston, St. Pancras and Kings Cross Stations are interesting features in their own right and continue to play an important role in the urban landscape. Our walk along Euston Road will include station visits, as well as a diversion to see Woburn Walk.

Euston Station

This station is set back across Euston Square, and the trees do their best to hide a mundane series of office blocks, a bus terminal and litter-strewn pedestrian areas. The redevelopment was part of the high water mark of urban renewal in the 1960s: the historic station with its famous Doric Arch was totally demolished, in spite of considerable protest, at the time of the electrification of the West Coast main line from London to Glasgow. Overhead electrical equipment meant a new train-shed, and the opportunity was taken to improve all the other passenger facilities. In their place was built a nondescript piece of Modernist architecture, functionally correct, but cold and uninviting for people who have to spend any time there, and something that is never likely to make it into the art/architecture books. It contrasts quite markedly with station redevelopments before or since. In a more heritage-conscious age a similar fate is unlikely to happen to other large stations, although having said that, Broad Street Station, next to Liverpool Street Station, was demolished in the late 1980s to make way for the Broadgate development.

(Return to Euston Road, cross into Upper Woburn Place and left into Woburn Walk, left into Dukes Road, right on Euston Road to St. Pancras and Kings Cross Stations)

Woburn Walk

This short pedestrian street, which is so easily missed, is one of the few examples remaining in Britain of purpose-built, Regency shops, dating from 1822 and built by Thomas Cubitt. It was once perhaps a local neighbourhood shopping street, but now contains an eclectic mix of functions, most of which are non-retail.

St. Pancras Station

If Euston was the high-water mark of post-war urban renewal, then St. Pancras performed the same role for 19th-century neo-Gothic. This imposing, red-brick development facing the Euston Road was the London terminal for the former Midland Railway which served the East Midlands and the North of England. It was designed by Sir Gilbert Scott and completed in 1865 as a single, high-arched span across all the platforms; the hotel which fronts the train shed was completed in 1874. The hotel has long since closed, offices have been relocated and the buildings for years were in a state of decay. Moreover, railway rationalization has resulted in this station being heavily underused; in 1988 all the suburban services were transferred to the north-south line (through Kings Cross Thameslink Station) to connect with various suburban services in south London.

The future for St. Pancras is uncertain. The buildings themselves are protected, and have been recently restored. The future could be assured if, as proposed, the station is tied into the Channel Tunnel high-speed rail link. The British Government's and British Rail's planning for an improved link to the Tunnel has become a national joke, especially when pitted against the already-completed French rail

link between Paris and the Tunnel. The latest scheme is for a route
through Kent, crossing the River Thames to the east of London, and
then proceeding on new or existing track to Stratford in east London and
St. Pancras. From here trains can connect to other parts of Britain. No
date has been set on this vital link from the Tunnel to London, and the
financing of the whole operation is in doubt because of a lack of
commitment on the part of the British Government.

Kings Cross Station

The next door neighbour, Kings Cross Station, could be thought of
as the dumpy poor relation. This much larger station, again with its
hotel (alongside rather than in front of the train shed), was the London
terminal for the Great Northern Railway, serving the East Coast route
to Scotland. The large number of main-line and suburban train
passengers over the years have put pressure on the facilities, the result
being an impressive facade obscured by more modern buildings which
are still insufficient for handling peak crowds.

Kings Cross has also been caught up in the Channel Tunnel plans:
an early proposal would have seen the demolition of the railway hotel
and a new Channel Tunnel terminal in its place. More important has
been the uncertainty, secrecy and outright bungling associated with the
reuse of the various railway lands to the north of the station. The
different proposals focus on market-driven office, retail and tourist
development, but overtaken by an uncertain economic climate and an
excess of office space. Opposition has been acute because the proposals
have little or no relevance to community interests and the desperate need
for community improvement in adjacent areas of St. Pancras and
Islington.

Kings Cross-St. Pancras area

Kings Cross-St. Pancras remains a focal point for transportation
routes; besides the two main-line rail terminals (one with suburban
services) and the Kings Cross Thameslink suburban service, there are
six London Underground lines, local and inter-city bus routes, and the
prospect of Channel Tunnel interchange facilities. However, the area
about two stations does not really reflect this high degree of
accessibility, suggesting that the transportation role of Kings Cross is
interchange rather than destination. The commercial services are of
little consequence, with small hotels and shops, for example, geared to
rail and bus passengers. There is also a small, and declining, residential
population to be found east and south of the stations which will

generate a need for local services. With the Town Hall, it is the administrative centre for the London Borough of Camden. Environmental quality certainly needs attention, including large volumes of road traffic (this inner-ring road mentioned earlier), a deteriorating housing stock, and a lack of greenspace.

> *(Return along the south side of Euston Road past the Town Hall, turn left on Judd Street, right on Leigh Street, left on Marchmont Street, right on Coram Street, right on Herbrand Street, left on Tavistock Place to Tavistock Square)*

Our walk in the area to the south of Kings Cross-St. Pancras Stations will take us through a relatively poor inner-London residential area. It was first developed in the late 18th, early 19th centuries with streets, squares and crescents containing terrace housing, although not on the same impressive scale as the area of Bloomsbury to the west. Some of this early housing still remains. Over time housing quality has declined and more recently it has been replaced by blocks of private and public apartments, many of which are now requiring renovation and upgrading.

Various institutional uses and other services have invaded the area, in particular small hotels and bed and breakfast accommodation, reflecting the close proximity of good transportation facilities and the various tourist sites in central London. Student accommodation for the University of London and its Colleges, either purpose-built or converted housing, is also found in this area.

Marchmont Street

The local neighbourhood shopping area found along this street and in the modern off-street shopping mall (Brunswick Shopping Centre) reflects this diverse local population. For example, the supermarket and other convenience shopping are for local residents, whether permanent or temporary, while the many restaurants are aimed at the tourist and student populations.

Peabody Estates

On the right at the northern end of Herbrand Street can be seen one of the estates of the Peabody Trust. Peabody was a philanthropist who in the late 19th century, in an era before largescale public housing, set out to improve the living conditions of the working class. There are a number of similar blocks of flats in other parts of inner London by

Peabody and others. It is difficult to appreciate looking at this gaunt, grey and uninspiring environment, with scarcely a flower or shrub to be seen, that this type of housing marked an improvement. However, it did. The flats were modest and certainly cramped for the average family size of that day; but they had basic facilities (for example, running cold water and sewers), were dry, and had separate living and bedroom accommodation. In today's climate the flats are lacking many basic amenities such as central heating, elevator access, insulation, security systems, and child play areas, and at the present time extensive renovations are being carried out.

Further reading

Edwards, M. 1992. A Microcosm: Redevelopment Proposals at Kings Cross. In Thornley, A. ed. *The Crisis of London*. London: Routledge, 163-184.

Lawrence, H.W. 1993. The Greening of the Squares of London: Transformation of Urban Landscapes and Ideals. *Annals of the Association of American Geographers* 83: 90-118.

Summerson, J. 1962. *Georgian London*. rev. edn. Harmondsworth, UK: Penguin Books.

Chapter 6

The East End before 1920

The East End is a world away from the West End, as any player of Monopoly will tell you! As the middle and upper-classes started to expand north and west of the Cities of London and Westminster after 1650, there was a more constricted working-class expansion to the east and north-east of the City and on the south bank, either along the river or the major roads into the county of Kent. In the massive population increases of the 19th century the East End developed as far as 10 miles from the centre of London. Its older, inner part became the classic Victorian slum; the later, outer part was often a boring expanse of unvaried, grid-iron, terrace housing, built to standards laid down in the late Victorian Health Acts. After the First World War, and especially after the Second, government intervention resulted in urban renewal in the older areas and new and improved housing developments further out.

The East End is economically and socially far removed from London, the World City, and especially its recent manifestation in adjacent developments such as Canary Wharf. The East End could not be less glamorous. There is mile upon mile of terrace housing and more recent blocks of flats, and much of the industry and transportation facilities, such as docks and rail yards, lay derelict. The inner area has become one of Britain's poorest areas, resulting from declining employment prospects, a government policy which saw the better-off working class population dispersed to the outer suburbs, and increasing cut-backs in public expenditures in an area so dependent on them.

Apart from Docklands, the area to the east of the City attracts little attention, in spite of its many problems. It is rarely seen by the visitor or indeed by anyone who does not have business there (or who has to pass through). On the other hand, it remains an important aspect of London's development, and we will follow its various phases and examine a wide variety of issues and problems that exist today. The first part will encompass the area that was developed before 1920 (and subsequently redeveloped, especially since 1945) (Figures 6.1-6.2), while the second part (in Chapter 7 below) relates to outer suburban and New Town developments after 1920.

(From Shoreditch High Street at Old Street cross into Calvert Avenue to Arnold Circus, right on Navarre Street, left on Boundary Street, right on Redchurch Street, left on Shoreditch High Street, left on Commercial Street, left on Fournier Street, right on Brick Lane to Osborn Street and Whitechapel Road, cross Whitechapel Lane to Commercial Road)

Shoreditch-Whitechapel area

Prior to the 1600s, there had been some strip development outside the gates of the City of London along what is now Bishopsgate and Aldgate, but the area between the two began to infill after 1600 when Britain's importance as a maritime and colonial power began to increase and London's growth could no longer be contained in the City (Figure 2.1). A second aspect to this development was the exclusionary nature of the City itself and the fact that various peoples and trades had to set up in this Shoreditch-Whitechapel area instead.

The first important group was a second generation of Huguenot immigrants from France (the first had gone earlier to Soho) who established themselves in the silk trades, resulting in the area becoming one of London's rag-trade centres, a role that can still be found today. Sephardic Jews also came here after being expelled from Spain. They engaged in the rag trades as well and accrued considerable wealth, but they were to be constantly discriminated against; not only were they forced to live outside the City, but they could not participate in many of Britain's institutions, including its banks, guilds, clubs, royal society and educational establishments. However, Jews were to find considerable success in a parallel commercial world.

In the 1800s there was considerable migration of poor and persecuted Eastern European Jews into England, principally into the Whitechapel area of Stepney. There was never a ghetto in London in the legal sense of the word, but by 1900 numerous streets in this area had over 90 percent of their population with Jewish origins. As Jews became more wealthy and or more accepted by mainstream society, they moved out of the East End, the greatest concentration now being

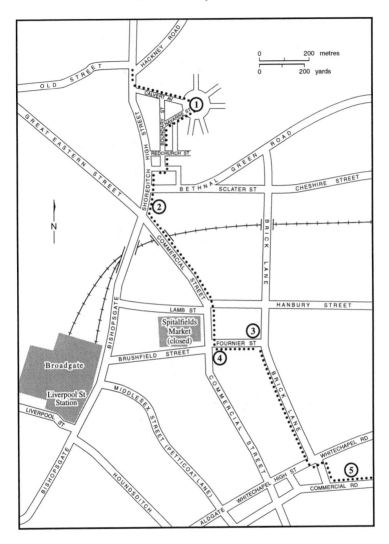

Legend

1   Arnold Circus
2   Bishopgate Station (closed)
3   Christchurch Spitalfields
4   Mosque
5   Bus to Limehouse Station

Figure 6.1   Shoreditch-Whitechapel Walk

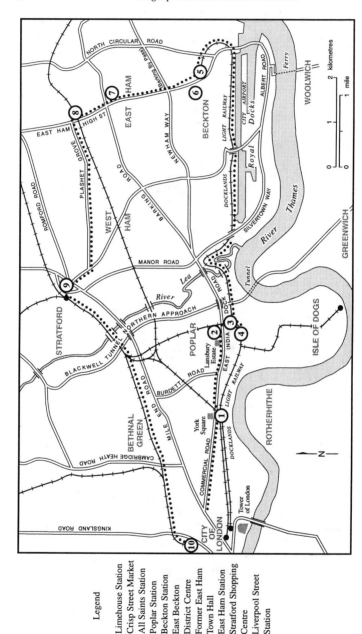

Figure 6.2 East End Excursion

Legend

1   Limehouse Station
2   Crisp Street Market
3   All Saints Station
4   Poplar Station
5   Beckton Station
6   East Beckton
    District Centre
7   Former East Ham
    Town Hall
8   East Ham Station
9   Stratford Shopping
    Centre
10 Liverpool Street
    Station

in Hendon in north London. Now that immigration of poor Jews has long ceased, the gateway factor of the East End and its large Jewish population are no more. Because of the institutions and social behaviour of Orthodox Jews, they are still spatially concentrated, but it certainly no longer compares with the East End at the turn of the century. Meanwhile, non-Orthodox Jews have less reason to be spatially concentrated.

Since the 1970s the Spitalfields-Whitechapel area has attracted a new rag-trade population, the Bangladeshis. They are facing especially difficult problems, coming from a rural area of Bengal and lacking in English language skills; on top of which they have deteriorating housing conditions, sweat-shop practices (mostly from employers from their own group), run-down services and racial prejudice to contend with. The mosque, one of which can be seen on Brick Lane, has replaced the synagogue in this part of London.

Boundary Street Estate

Over the last hundred years this working class area just outside the City of London has seen various schemes aimed at improving housing conditions. This particular example, in the Boundary Street area just off the Shoreditch High Street, was the London County Council's first large slum clearance scheme, completed in 1900 and housing some 5,000 people (Plate 6.1). The five-storey walk-up units were revolutionary in their day and designed to maximise on airiness. A garden and bandstand was provided, as well as a laundry, shops, schools and workshops for local entrepreneurs. The slum problem was hardly solved: more people were displaced than rehoused, and the rents could only be afforded by the better-off elements of the working class. Moreover, after 1900, the London County Council became more interested in two-storey terrace and semi-detached housing on the fringe of the built-up area. Time and neglect have certainly taken their toll on places like the Boundary Estate, and so much of the redevelopment needs as much attention as the remaining older housing stock.

Spitalfields

In addition to the rag trade, the other important role for this area, at least until the 1970s, was an inner London freight distribution centre. Liverpool Street is the terminus for the former Great Eastern Railway, opened in 1874. Before that, trains terminated at Bishopsgate Station, just east of Shoreditch High Street and Commercial Street, and until

Plate 6.1 Boundary Street Estate from Arnold Circus (Photo: London
Borough of Tower Hamlets)

recently this was a freight depot with extensive tracks stretching east
towards Bethnal Green.   The facilities lay derelict awaiting
redevelopment.

Nearby, although not directly connected, is Spitalfields Market, one
of the many London wholesale food markets noted earlier; it has been
relocated and the site also awaits redevelopment.   Likewise, many
associated servies have closed and some of the properties have been
taken over by an expanding rag trade.

Brick Lane

This is the focal point of the Bangladeshi community, its textile and
clothing industries and many other services.   The walk along Fournier
Street before this will give some impression of the area's early
development, including the Georgian-era housing and workshops and
the parish church of Christ Church Spitalfields, built by Nicholas
Hawksmoor in 1720 and intended to counteract the threat of Huguenot
Protestant dissenters.   The Huguenots built their own chapel at the
corner of Fournier Street and Brick Lane in the 1740s; but the changing
immigrant scene resulted in the Methodists taking over the building in

early 1800s, followed by Orthodox Jews in the 1890s and more recently the conversion of the synagogue into a mosque.

For years the area has been fraught with poverty, exploitation and racial tension, attracting in turn a wide range of social reform movements. Conditions were highly publicized at the time of the Jack the Ripper murders, and again more recently when racist attacks occurred in the area and poor living conditions were exposed.

(From Commercial Road take the outbound no.5,15,15B or 40 bus to Limehouse Station; from the Commercial Road turn left on Barnes Street to York Square, return to Commercial Road)

Limehouse: a working-class suburb

Such has been the transformation of the East End (and indeed other areas of inner London) since 1940 that it is now impossible to see a large expanse of the early 19th century working-class housing. The worst tenements and courts built before the 19th century have long gone. However, it is now exceedingly difficult to do likewise with the grid-iron roads of small, two-storey terrace houses that were built before the 1880s. After the London County Council was formed in 1889 it very much became public policy to start removing the worst of working-class housing; but the task was immense and by 1940 the terrace housing of places like Limehouse remained largely intact.

The blitz of 1940-1941 brought mass destruction to the houses of the East End. The number of destroyed and damaged houses, together with the evacuation of people to safer areas, resulted in considerable population losses. If there had not been the doubling up of families in the one house, losses would have been even greater. After 1945 programmes to replace these shoddily-built and poorly serviced housing areas now proceeded with a vengeance. The Labour government of the day felt it owed to it to these East Enders who had endured so much in the war and had put up with such bad housing for decades before that. Between 1945-70 what the Nazis didn't get, the public housing authority did! Little remains, but in the York Square area of Limehouse the terrace housing has been preserved and it is possible to reconstruct the community that once existed (Figure 6.2; Plate 6.2). The major components are as follows:

The House

It is strictly terrace housing here with passages at intervals through or between the houses so that access could be gained to backyards.

Plate 6.2   19th-century terrace housing in the York Square area of
Limehouse (Photo: London Borough of Tower Hamlets)

You may note that the housing is not uniform. Differences in size,
design and ornamentation were indicative of different status or income
levels in the working class community (e.g. skilled, semi-skilled and
unskilled workers, regularly versus irregularly employed etc.). Houses
would be of two types: 1) two-up (two bedrooms), three-down (front
parlour, living/dining room, with kitchen projecting into the yard); 2)
three-up (three bedrooms), three-down (same as 1 above) - this was
called the tunnel-back house and was the norm for working-class
housing between about 1875 and 1920. This arrangement, however,
would change if more than one household lived in a house. In 1931 a
nearby community (Bethnal Green) had 50 percent of its houses
containing two households; in 1951 it was still 25 percent. Prior to the
1950s probably, there was no bathroom, no indoor plumbing apart from
a cold-water tap in the kitchen, and the toilet would be outside either at
the end of the yard or adjoining the kitchen wall. The yard would
contain a coal-house, a washing line, perhaps a shed, rabbit hutches,
pigeon loft etc. There may be a small garden for flowers or vegetables.

The Street

The dominant plan in Limehouse and other inner areas was a grid-iron arrangement of narrow streets. Until the 1870s the building line would be the edge of the pavement, which itself was narrow. Prior to mass ownership of the car after the 1960s, these streets would have seen little traffic; therefore, their width was more a function of letting in sunlight, fresh-air and creating a sense of space for children to play and for the adults to interact. The depth between streets did not allow for very long yards Unlike the West End, Belgravia and Bloomsbury, only in rare instances (one being York Square) was this plan broken up by the square or some other form of open space. The lack of trees or other vegetation is still very obvious.

Until 1940 the population was quite immobile, with most people being born, married and growing old either on the same street or in the same locality. In a survey of an East End street of 59 households, 38 of those were related to one or more of the remaining 58 (Young and Willmott, 1962, 41). The key to this relationship was mother and married daughter who wished to live close together; thus, in the era of private rental it was important to be on good terms with the rent collector who would look after housing vacancies. There was intense loyalty to the street; they were not just your neighbours but your relatives and support system. At times of celebration (a Coronation, Jubilee or war victory) street parties would be held.

The Pub

At street corners there were often one or more pubs where adults would socialize. (York Square still has two.) It was often called your 'local', and you were intensely loyal to the one pub, its brewery and beer and the publican who ran it. In working-class communities only close relatives socialized in one another's houses, and even then it may have been only on special occasions such as weddings and funerals. Houses, it must be remembered, were very small and already desperately overcrowded. The more spacious meeting place was (and is!) the pub. It is a peculiarly British institution that has not been well replicated in other countries; until recently it was usually socially segregated into a noisy public bar, where the drinks were cheaper and the furnishings more spartan, and a quieter lounge or saloon bar where the clientele would be more 'refined'. It was strictly a place to drink, not to eat; the working classes rarely ate out, except when on holiday. It was in some ways a social and sports club (soccer and dart teams, summer outings to the seaside) and the focus of a community support group in the era before the Welfare State. Pubs were open all day, but from the First World War until the 1980s hours were restricted in the hope of curbing drunkeness and family debt. The pub was considered a man's world. A woman alone at the pub could easily be regarded as a prostitute; even two or more women socializing together would do nothing to enhance reputations! When urban renewal swept away the terrace housing, the pub was one of the few buildings to remain.

Shopping

Grocery shops no larger perhaps than the front room of the average house were found throughout these inner-city areas, particularly on street corners; and this was where women would both socialize and purchase everyday needs. Some women might go further afield to do a weekly shop at a larger (Co-op) store where prices were better, but groceries had to be carried home. Purchasing every day was more convenient to some, an opportunity to get out of the home and socialize, and a key to obtaining credit until money could be found to pay the bill. The corner store contained mostly packaged and canned groceries, dairy products (not milk, which was always delivered), some baked goods (bread was usually delivered) and a limited range of fruit and vegetables, meats, hardware items, and perhaps newspapers, magazines, tobacco products, and a sub post-office. Urban renewal and changing shopping habits have seen the almost complete demise of the corner store.

Shopping centres would be lines of shops on main roads such as the Commercial Road, and would include shopping, as well as convenience, goods. Shopping for the working class was very localized; higher-order shopping in regional centres was poorly developed until after the Second World War. The lines of small shops have again made way for improved shopping in district and regional centres; those that have survived have often been converted to other uses. In the present era of heavy traffic on main roads, lines of shops do not make for a pleasant shopping environment; Commercial Road can certainly attest to that.

Shopkeepers, along with publicans, teachers and clergy, were the backbone of the small, middle-class population resident in communities such as Limehouse.

The School

A school became a feature of every community in the last half of the 19th century as universal, free education became mandatory for children between the ages of five and twelve, and gradually extended this century to sixteen. A school would serve a large number of streets, or a neighbourhood. As the age for leaving school advanced, education was divided into primary and secondary schools; the latter were invariably larger and served a wider area. Schools, like pubs, have frequently survived urban renewal (because housing was a different department of the local council from education!) and often appear as gaunt, grimy, two to three storey buildings within their hard-surface play areas. Compared to schools in middle-class or newer working-class areas, they often have poorer facilities to match their lower levels of attainment.

Other community institutions

In a very paternalistic way, the plight of these desperately poor working-class, inner-London communities such as Limehouse attracted a variety of middle-class or establishment attention which would be seen in a range of functions and buildings. The last decade of the 19th century saw the establishment of effective local government in the new County of London (LCC) and the various Metropolitan Boroughs. Limehouse became part of the Met. Borough of Stepney. (In 1965, Stepney was to join with Poplar and Bethnal Green to become the London Borough of Tower Hamlets.) The Town Hall and the local offices of a county department provided valuable and growing services to the community and were an important source of regular employment. Another service, the Municipal Baths (now defunct), were built to overcome the absence of bathrooms and running hot water in homes; they were not places to swim! Similar facilities were provided to do the weekly washing.

The Church, usually the Church of England, but sometimes the Catholic Church, the Chapel (Methodist, Baptist and Congregationalist) or the Synagogue, was built at the same time as the estates of houses. Churchgoing itself was certainly in decline, but a large amount of missionary work was done with the local population. Settlements, orphanages, missions, friendly societies and clubs, some of them residential, abounded in these areas; they were charitable organisations with a mandate to help specific groups, often with volunteer help provided by London's middle and upper classes.

As universal suffrage was extended to include working class men in the late 19th century, the political parties began to extend their influence into areas such as Limehouse. The Conservative party was only ever of marginal importance; these areas were first Liberal and after 1900 increasingly Labour, reflecting a broad mandate to help the working classes. Local government in urban Britain has always been very party-oriented, and as Labour took over local councils and thus the Town Halls, it deliberately set out to help its working-class supporters (sometimes fi.nding itself on the wrong side of the law laid down by Parliament) These councils over the years took on more and more services, something that was not reversed until the 1980s.

Trade unions of blue-collar workers in different industries and trades organised in order to improve their wages and conditions. Prior to the Welfare State, unions had an important support role; they also used it to promote and finance the Labour Party at the local, county or national government level.

Industries

Outside of Docklands and the heavy industries of Thameside and the Lea Valley, there was a very diversified, small-scale industrial development interspersed thoughout these older residential areas of the East End, especially textiles and clothing, furniture making, footwear, and small metal goods. It was similar in age and scope to those older industrial areas seen in the West End. There has been a wholesale decline of these industries today, but a hundred years ago they were the backbone to the community and provided a great deal of stability to it. Wages were low, but for many they did come regularly. However, working conditions and hours of work were usually horrendous. In the clothing and footwear industries work could be taken home, and whilst it was mostly the responsibility of the woman, it could be a family affair.

Distributive trades

So much of what the community needed had to be brought in and then distributed. Transportation services and the goods themselves employed large numbers of people. The railway, the canal and the docks were the principal modes; it would then be off-loaded on to an army of horses and carts for distribution. Feeding and caring for horses was a major occupation in itself. Coal was used for household and industrial heating, and for steam generation for power in industries and on the railways; thus, rail yards, coal depots, coal merchants etc. were to be found everywhere. The advent of motorized road transport, increased market size and the demise of coal have changed the nature of distribution; homes are heated by natural gas, goods are delivered to shops from hundreds of miles away, and horses and railway yards are now oddities.

(From the Commercial Road at Limehouse Station take the outbound no.5,15,15B or 40 bus along East India Dock Road to the Crisp Street Market)

Redevelopment of inner London

The Crisp Street Market and adjacent Lansbury Estate in Poplar afford us an opportunity to examine the massive redevelopment of the East End. The blitz of 1940-1941 and the urban renewal projects of the LCC, its successor the Greater London Council (GLC), as well as local councils, did more than clear away acres of mean streets and poor housing: it wrecked for ever a well integrated working-class society. The intention of government was to provide better quality housing than the private entrepreneur had done nearly a hundred years earlier, to reduce overcrowding by assigning one household per house or apartment, to replace damaged and destroyed properties and bring back people who may have had to leave the area, and to provide a better housing and community environment. In places such as Stepney and Poplar the vast majority of new housing thus became publicly owned.

Slum clearance had been going on since the last half of the 19th century, much of it by charitable trusts (the Peabody Estates, for example, seen earlier), rehousing people in four to seven storey (walk-up) apartment blocks. After 1889 the LCC took over this role, but it was hardly a dramatic change in the older parts of the East End until after 1945. The advent of a Labour government at Westminster resulted in more money being set aside for urban renewal, improved legislation to facilitate change, and a redesignation of what constituted a slum house that was unfit for human habitation. In many instances there was nothing much wrong with the house that could not have been taken care of by normal maintenance (and landlords had neglected this over the years), rewiring, inside plumbing, a new roof etc., (features that can be seen in the restored housing around York Square.) However, governments in their wisdom believed new housing to be better and cheaper, and cleared urban spaces would provide for all the other services that were needed (for example, public open space, new parades of shops, the health clinic, soccer pitches etc.). Remember that the blitz had already done a very good job of removing the old housing stock and upsetting community relations.

New housing facilities in the East End were very rarely modelled on the old. Between 1945 and 1970 the narrow grid-iron streets with their terrace housing were torn down, and plans were drawn up for large new estates where the arrangement of the homes bore no resemblance most of the time to the former streets. Few families any longer lived in two-storey houses; most were accommodated in multi-storey apartment blocks arranged in pedestrian-only precincts, surrounded by ample amounts of landscaped public open space and children's playgrounds. All the homes came with better amenities (for example, bathrooms and running hot-water, even central heating). Although the stamp of the LCC's Architect's Department was seen on all these projects, some were by award-winning architects and most were hailed as giving people who deserved it a step-up in life. A new utopia was rising on the slums and bomb sites of the East End, although the cynics were already saying that dreary apartment blocks were replacing dreary houses.

So extensive was the renewal after 1945 that it in fact became recognised as the Stepney-Poplar Comprehensive Development Area, the boundaries of which were the Whitechapel and Mile End Roads in the north, the Commercial and East India Dock Roads in the south with Burdett Road being the north-south spine road. A 'New Town' for approximately 100,000 people would be constructed in stages by the LCC and the Boroughs of Stepney and Poplar. Whilst this reconstruction has now been carried out, the lasting effects are certainly not exhilerating. Decline has long since set in and further renewal is necessary.

Lansbury Estate

Our excursion now takes us to one of the earliest of these renewal areas. The Lansbury Estate is located west of the Crisp Street Market as far as Canton Street, and between East India Dock Road and Cordelia Street (Figure 6.2). This development was associated with the Festival of Britain in 1951. The intent was to replicate the Great Exhibition of 1851, and show Britain's many achievements over the intervening hundred years, as well as give a boost to British morale in the immediate post-war years of austerity, shortages and rationing. The main exhibition site was on the South Bank near Waterloo Bridge, but other locations, such as the 30 acre Lansbury Estate (part of a 124 acre neighbourhood), were used for specific purposes, in this case planning and architecture.

This urban utopia for the working classes was hardly revolutionary, but for inner-London the LCC did try to promote a compact and neighbourly community of new two-storey housing, and some low-rise apartment blocks, in traditional London brick and slate, on traditional streets and at a human scale (Plate 6.3). The adjacent Crisp Street Market was the first attempt in London to create an environmentally-improved, pedestrianized, district shopping centre away from the main

Plate 6.3 Lansbury Estate. Much of this two-storey housing is now in private ownership and the uniformity of the original 1940's house has been lost through individual improvements to property. In the background is the Canary Wharf tower (Photo: London Borough of Tower Hamlets)

road. The then Borough of Poplar commissioned a clock tower at the corner of the market to convey old-style civic pride, as well as provide a focus and a sense of height to a two or three storey world.

(Walk along Market Way, turn left through the shops into Ricardo Street, left on Bygrove Street, left on Grundy Street and return to the Crisp Street Market)

The earlier plan for the Stepney-Poplar area of 11 neighbourhoods like Lansbury, each with their community facilities and a measure of self-containment for approximately 10,000 people, was not adhered to. Subsequent development was unfortunately piecemeal and has been in the form of low-rise apartment blocks. In the 1960s these gave way to more aggressive designs, in particular the tower blocks (high rises) and apartment complexes connected by high level walkways. These new estates allowed for even more public, off-street spaces, and can be seen in the vicinity of the Lansbury Estate (Plate 6.4). However, it was

Plate 6.4   Low and high-rise apartment development in the Lansbury
area (Photo: London Borough of Tower Hamlets)

increasingly found that imaginative designs and the best of housing
facilities that the public sector could afford did not make for good
living environments. The public-sector housing industry was literally
to come crashing down like a pack of cards: a fatal gas explosion in
1968 in the kitchen of a high-rise apartment in the London Borough of
Newham resulted in the destruction of all the kitchens above and
below.

Troubles in Utopia

Some of London's greatest levels of deprivation are to be found in
these redeveloped areas. We will examine why this post-war utopia has
foundered so badly; some of the evidence can be seen on our East End
excursion:-

1) *Emphasis on improved housing rather than community integrity*
Central to the redevelopment idea was the provision of better housing.
There is no doubt that this was done. In new public housing, as
opposed to old private rental housing, families had more space (often
because families were now smaller!) and better facilities. However, the

community as it existed was destroyed. When people were moved and rehoused, they were assigned housing on an individual family basis. The old street was broken up and scattered all over the new housing project, while some families were assigned housing in out-county and New Town estates miles away and a few better-off people moved to private housing areas elsewhere. The relationship between mother and daughter was often forcibly ended, since nepotism was not allowed by the public authority in the assigning of housing. The corner-shop, even old shopping strips, were demolished, although pubs and schools tended to remain.

2) *The deterioration of housing quality* While the new housing in the East End certainly had better facilities, the overall quality of the design, construction, building materials and workmanship frequently left much to be desired. Technological inadequacies, short-cuts and profiteering, cut backs and saving tax monies all entered the picture. Twenty years or less after the housing was built some of it became uninhabitable. The high-rise gas explosion was the catalyst for a change in thinking.

Much of the deterioration was helped along by vandalism and misuse. For example, living in a tower block became difficult when elevators were out of action so much of the time; it was as much people abusing the equipment as it was mechanical problems with the equipment per se.

3) *Problems of housing design* Stacking people up in the sky, the creation of anonymous public open spaces, the maze of pedestrianized areas, the loss of privacy and individual space and the poor integration with community services were to have devastating effects. Children's play areas were much more divorced from where their families lived (i.e. many floors up!); it was more difficult for people to interact since the block did not mirror image the street; people could no longer watch out for one another and their homes in the way they used to; the myriad number of public thoroughfares invited strangers into an area, made them difficult to police and encouraged petty crime; families no longer had a garden, however small, and the balcony did not give the same sense of private open space, a place to park the bicycle or keep the rabbits. People increasingly objected to their living spaces and began to take out their frustrations on that space.

4) *The ghettoization of the poor* Public housing has long had its own social stratification. The best and most preferred housing is assigned to the best families (where employment and income are

regular, there is no history of non-payment of rent, no property damage etc.), while problem families of various kinds are kept together by the housing authorities in the poorest of housing projects. This public policy, which is more often covert than overt, comes on the back of dispersal policies which have resulted in a disproportionate share of the skilled, younger and better able working class families moving out of the East End altogether.

5) *The collapse of the industrial economy* Deteriorating housing quality and worsening social conditions are a reflection of the collapse of the employment base on which the working class in the East End depended, including the closure of the docks and the loss of industry and dock trades, and the lack of suitable alternatives in the vicinity. This was not something unique to London, but it seemed that areas with high percentages of people in manufacturing (as some parts of London were) suffered higher rates of job loss. Unemployment rose sharply; it can be nearly 25 percent in some inner areas whereas nearby outer areas may have only 5 percent. The lack of education and skills amongst the population detracts from various kinds of employment coming here or commuting a mile or so to jobs in the City or Docklands. A deprived population is further deprived.

6) *The slowing down of public spending* The 1980s brought about a change in government policy. An area that was so dependent on public monies found itself subject to decreasing government expenditure. Also, there was a political gulf between free-wheeling Labour local councils and the Conservative administration in Westminster. There were enormous cut-backs in the public house-building program at a time when there was a shortage of affordable housing in the London area and a boom in the housing market. Selling off public housing was forced on local councils, thereby accentuating the housing problem.

7) *Changing composition of the population* The deprived nature of the population who remained in the East End has been aggravated by its changing composition. The industries of the area have long been attractive to foreign immigrants. After 1945 there was a flood of immigrants from Britain's colonies in the Caribbean, south Asia and parts of Africa, reflecting a booming British economy (and a large number of unskilled and semi-skilled blue-collar jobs that British people did not want) and few, if any, restrictions on immigration. Many of these immigrant groups can be found in the East End.

Since the 1960s Britain has continually placed restrictions on new immigrants, in part racially motivated, but also in response to rising unemployment and drastic reductions in labour needs in the blue-collar trades. However, natural increase amongst the existing immigrant population means that younger generations (increasingly born in Britain) are still caught up in the problems of falling labour demand and poor educational and work skills preventing shifts to other employment sectors. Sadly, non-Whites have poorer educational skills, poorer housing, more health problems, less income and more likely to be unemployed compared to Whites in the same area. This deprivation has fuelled resentment on the part of non-Whites, and violence rather than sympathy between them and various White groups.

The redevelopment of the redeveloped areas

Drastic social and economic changes after the 1960s in areas such as this in the East End, accompanied by urban riots, have resulted in the government paying more attention to inner city problems. Studies have been commissioned, results published, legislation enacted and public monies committed to deal with the problems of deprivation. By necessity it has to be a multi-faceted, but co-ordinated, attack dealing with education and retraining, jobs, income, housing and community services, which is something that did not happen in the utopia of the immediate post-war period.

In housing we shall see that there has been a very active program to improve the public-housing estates, even those built in the 1960s and 1970s. The demolition of tower blocks is most noteworthy. Second, there is the conversion of many low-rise apartment blocks into townhouses, taking off the upper floors where necessary, adding front/rear doors, gardens, garages etc, and rearranging internal spaces. Third, the wholesale improvement of services both internal and external, either replacing shoddy work of old or bringing services up to modern standards, such as adding central heating, elevators and double-glazed windows. Fourth, the reduction of public spaces and the corresponding expansion of private spaces, important factors at a time when increases in crime have been considerable.

(From Crisp Street Market cross East India Dock Road to All Saints Station, take southbound Docklands Light Railway (DLR) train to Poplar, change on to Beckton line to Beckton Station and the East Beckton District Centre))

Beckton

The present A13 East Ham-Barking By Pass once marked the southern limit of the built-up area of East Ham. Beyond that was a vast tract of marshland which contained the last of the Royal Docks (opened in the 1920s), the small dockland communities of Silvertown and North Woolwich, extensive sewage treatment facilities (at the end of the Northern Outfall which serves a large area of north and east London), gasworks, industry and vacant land. The closing of the docks resulted in this whole area changing in character. Our field excursion into Docklands (see Chapter 8 below) will look at whole political economy of this change, but at this point it is appropriate to examine one of these new Dockland communities.

Beckton, lying between the A13 road and the former Royal Albert Dock, and stretching east-west between the older suburb of Custom House and the gas and sewage works, was opened up by the London Docklands Development Corporation (LDDC) in the 1980s for private housing, and almost entirely consisting of two-storey terrace and semi-detached homes. As in the former Surrey Commercial Docks on the south side of the River Thames, these extensive back estates, well away from river, or even dock, frontage contain much more affordable housing (although still out of reach for most of the East End population).

The area was also provided with a range of community facilities, parkland, improved shopping in the form of the East Beckton District Centre, a retail park and second superstore, an industrial park, business centre, and various sport facilities including golf courses and the Beckton Alps Ski Centre! Improved road communications now link North Woolwich and the river ferry with the southerly extension of the North Circular Road, and a new East London River Crossing is proposed. Meanwhile the DLR has been extended from Poplar along the northern side of the Royal Docks to the East Beckton District Centre; unfortunately, the new line crossing derelict land symbolizes so well the lack of integration between transportation and land-use planning.

(From East Beckton District Centre take no.101 or 300 bus to Barking Road, walk along High Street North to East Ham Station)

Housing reform in inner London after the 1870s

The incredible growth of London in the 19th century, its continuing high population densities and poor housing quality, especially in more industrial, working-class areas, led to two types of reform movement:

The Garden City Movement

A small but very significant shift in the nature of urban development in the last half of the 19th century saw various reformers, many of whom were industrialists, philanthropists and planners, turn their backs on the industrial city in effect and provide people with a better living environment beyond the city on a green-field site. It was not simply to be another housing estate, but an integrated community with housing, shops, parks, community services, industry and other employment with good transportation to nearby cities. Ebenezer Howard's Letchworth Garden City (1903) is the most well known in the London area; a second Garden City at Welwyn (1922) later became one of the London New Towns. From these small beginnings came an important change in British urban planning after 1920 which we will trace in a later field excursion (see Chapter 7 below).

Urban planning and the Health Acts

By far the most important aspect of urban change in the late 19th century was related to the various pieces of government legislation dealing with health and local government reform. The inadequate nature of sewage and water supply and the occurrence of typhoid and cholera epidemics eventually led to extensive reservoirs, water purification plants and mains water to most London homes, and the separation of this from a sewer system to each home and treatment plants (sewage farms) on the marshes on Lower Thameside.

Government legislation on health was amongst the earliest planning legislation, since house building was regulated in ways that had not existed before. These various Health Acts allowed local government to set building lines and minimum distances between houses on either side of the street and between streets, in effect lowering densities. Given that working-class housing was terrace housing, densities were still high by modern standards (approx. 50 per acre), but the overall appearance was one of considerable contrast with earlier development. Roads were longer and wider, houses slightly larger (the tunnel-back house was still the most common), front 'gardens' universal (and thus the introduction of some form of greenery into the built environment), and back gardens which were large enough to include vegetable plots. We can see these types of housing and streets off High Street North in East Ham.

These so-called by-law housing areas of inner-London were very extensive, and although not expressly forbidden (as it would have been under later planning legislation) did not contain large, obnoxious industries. A combination of actors, including the speculative house-building industry, the railway companies and the needs of industrialists resulted in this increasing separation of residential and non-residential development.

The role played by the railway companies cannot be underestimated. The housing could not have been sold and then rented if the railway companies were not able to transport workers daily to their employment in, or closer to, the centre of London. On the other hand, commuter fares were an important source of income for the companies, most of which were national or regional in their coverage with extensive mileage of lines in low populated rural areas. Decentralization of population by the railway companies was first an upper-class and then middle-class preserve, but working-class housing was not far behind, and certainly did not have to await the arrival of cheap electric street-car (tram) transportation at the turn of the century. (Indeed, in the London context the tram was slower than the train and was used more for local journeys; trams only replaced trains and stations within about two miles of the London terminal.)

Acts of Parliament could compel railway companies to offer certain fares on certain trains; and one reason for the extensive areas of working-class housing to the east and north-east of central London was that the Great Eastern Railway Company out of Liverpool Street Station had to offer extraordinary low fares on some of its trains, something that no other company was compelled to do to the same extent. We are going to examine one of these working-class suburbs that is served in part by the Great Eastern Railway.

East Ham: the Victorian and Edwardian suburb

East Ham lies to the north of the Royal Docks complex and is served by three important railway lines: the former Great Eastern (GE) suburban lines out of Liverpool Street Station through Manor Park, the former London, Tilbury and Southend Railway (LTS) out of Fenchurch Street through East Ham itself, and the parallel tracks of the District Railway (now the District Line of the London Underground).

East Ham grew principally in the years immediately before the First World War. The population in 1871 was 9,713 and by 1901 it had grown to 69,758 and by 1921 to 143,246. Such was the importance of East Ham and the neighbouring inner London borough of West Ham (since 1965 amalgamated to form the London Borough of Newham) that they resisted being included into the LCC area in 1889. Indeed, no part of the County of Essex was included in spite of the fact that there were extensive suburban developments by this time in what are now the London Boroughs of Newham, Waltham Forest, Barking and Dagenham and Redbridge.

East Ham was a very small village before the 19th century and few traces of it now remain. Its public and commercial buildings lying along the old A12 or A13 roads out of London, and the High Street between the two, exemplify the late Victorian, neo-Gothic period of development.. They exude a period of relative prosperity in the industrial life of London. The Town Hall, perhaps the borough's only memorable building, is an architectural gem! After the meanness of the streets of Stepney and Limehouse, East Ham's houses represented the respectable lower-middle and working-classes - the folk who had arrived, who had the more skilled jobs and who could afford to pay the much higher rents and the extra costs of commuting.

Housing refurbishment

The urban renewal promoters of the immediate post-war period never really reached areas like East Ham. They were too busy demolishing the terrace housing closer to central London; and by the time they reached East Ham, government policy had changed. East Ham's turn of the century working-class housing, which we can see in the roads on either side of the High Street, has become part of the post-1970s era of refurbishment of the existing housing stock. It was increasingly realised that refurbishment was cheaper than renewal, did not upset existing community relationships, encouraged home ownership and better protected private spaces. The one very obvious disadvantage is that these early suburbs can in no way accommodate the car owner; streets are jammed all day with parked vehicles and are hazardous to life and limb, especially as residential streets are so often used by people trying to avoid traffic jams on the main roads.

East Ham, like so many inner-London suburbs, has recently attracted a new immigrant population. The principally south Asian group here has been attracted by the ability to own their own homes and taken advantage of the long-standing, good race-relations in this part of London to introduce their own institutions. This has very much changed the character of business on the High Street where so many shops and offices are now geared to an Asian population.

*(From East Ham High Street/Plashet Grove take a no.104 or
238 bus to Stratford Broadway)*

Stratford

Stratford is the regional centre for inner east London, with a
shopping centre focusing on the Broadway, the historic core of the
small town that was originally here. It became the administrative
centre of West Ham (now part of the London Borough of Newham), a
largely 19th-century working-class borough with an extensive area of
Dockland and industries on its west side along the Lea Valley.
Stratford was also the rail centre for the Great Eastern Railway,
developing on the fringe of the built-up area in the 1830s. It contained
British Rail's largest locomotive depot until the end of the steam era.
There is still a diesel locomotive works, plus extensive carriage sidings
and freight depots. However, rationalization as well as industrial
decline have resulted in vast tracts of derelict or underused land, and as
a result there is a proposal for Stratford being a London freight terminal
for the high-speed rail line connecting the Channel Tunnel with St.
Pancras Station.

Stratford is a very important rail and bus interchange point between
the Central Line of the Underground, the British Rail inner and outer
suburban lines from Liverpool Street, British Rail's North London line
from North Woolwich to Richmond (one of the few ring lines!), the
DLR branch from Canary Wharf and numerous local bus routes. The
completion of the Jubilee Line from central London via Docklands to
Stratford will enhance communications south of the River.

The main street, the Broadway, was where the A11 and A12 routes
diverged. The old parish church formed an island in the road, and the
broad street contained a market. The forces of urban renewal, the need
for improved transport connections and better shopping facilities
resulted in a comprehensive redevelopment programme in the 1960s.
An inner ring road was completed around a new enclosed shopping
mall, which connected to the station, bus park and multi-storey car
parks. Thirty years on, the shopping centre is in need of revitalization.

*(Return to central London via British Rail or London
Underground Central Line from Stratford Station to Liverpool
Street)*

Further reading

Brownhill, S. and Sharp, C. 1992. London's Housing Crisis. In Thornley, A. ed. *The Crisis of London*. London: Routledge, 10-24.

Cross, M. 1992. Race and Ethnicity. In Thornley, A. ed. *The Crisis of London*. London: Routledge, 103-118.

Harrison, P. 1992. *Inside the Inner City: Life under the Cutting Edge*. Harmondsworth, UK: Penguin Books.

Young, M. and Willmott, P. 1962. *Family and Kinship in East London*. Harmondsworth, UK: Penguin Books.

Chapter 7

The East End after 1920

The second day of our East End transect takes us into an outer suburban area that was developed after the First World War, the Green Belt beyond that and a post-1945 London New Town in the county of Essex The overriding theme is still the outgrowth of a working class community, and in particular the importance of public-sector housing, built as large estates in park-like settings on the urban fringe. Calling such areas the East End could be regarded as geographic licence, even offensive since so many Londoners regarded migrating out to Essex as a means of escaping the East End. However, the strong social and economic ties between inner and outer areas underlie this consideration of an extended East End.

By 1920 urban growth in London had reached roughly 8 to 10 miles from the centre of the city. The limit was denoted by the long lines of terrace housing which until that time constituted the vast majority of both working class and middle class housing. After 1920 housing development was to change drastically. Terraces were rarely more than eight houses and usually four or six. More common in fact was the semi-detached house, especially for middle class housing. (The middle class suburb will be viewed in Chapter 9 below.)

On the east side of London, Wanstead, Ilford and Barking were the limits of the built-up area in 1920, with some small developments beyond this clustering around the stations along the former Great Eastern Railway, for example Chadwell Heath and Romford (the latter was a small Essex market town at this time). During the next 20 years urban development was to extend over nearly all of Wanstead-Woodford, Ilford, Barking, Dagenham, Romford and Upminster-Hornchurch. It was principally private, semi-detached housing for middle-class and better-off working class families. However, in the midst of this inter-war development the LCC built the largest public housing estate to date at Becontree on the Fenchurch Street to Southend railway line. The first part of today's excursion will examine this very significant urban development.

The continuous outward growth of London was stopped in 1939 by the Second World War and the institution of new planning controls, in particular the establishment of a Green Belt around London. This forced the vast majority of new post-war developments (as opposed to the redevelopments seen earlier) to leap-frog the Green Belt to the towns and villages beyond. Again, the lion's share of these post-war developments was private housing; these were built at increasing distances from central London (today up to 100 miles!) and were predicated on land availability, transport improvements, life-style changes and housing and commuting costs. An important departure from this private-sector development, and the continuing LCC (later GLC) public housing, was the promotion by central government of the New Town idea and the development initially of eight New Towns around London. The second part of today's excursion will look at one of these New Towns at Basildon in Essex.

(From central London take a British Rail suburban train from Liverpool Street Station to Ilford; or take the London Underground Central Line to Stratford and change to a British Rail train to Ilford)

Ilford: a London Regional Centre

For a Greater London of over 6 million people a disproportionate share of the high-order retailing and other service functions has been concentrated in the City of London and the West End. Since Abercrombie's 1944 Plan and the Greater London Plan of the 1960s there have been better attempts to break this log jam, and a number of older town centres have been expanded with higher-order services, including Romford and Ilford in the east, Wood Green and Brent Cross (a new North-American style centre) in the north, Ealing and Hounslow in the west, and Kingston, Croydon, Bromley and Lewisham in the south.

The outset of our excursion today will take us to one of these large centres, Ilford (Figure 7.1). The town grew mostly in the late 19th century, a more middle-class version of East Ham (seen in Chapter 6 above). Its High Road was the main road (A12) out of London, in fact the old Roman Road from London to Colchester. As is the case with

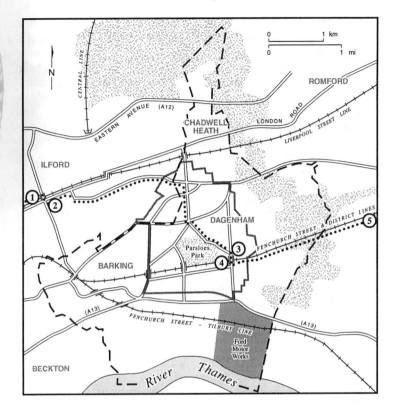

Legend

1	Ilford Station
2	Ilford Town Centre
3	Becontree Estate Walk
4	Dagenham Heathway Station
5	District Line to Upminster
— —	London Borough of Barking and Dagenham boundary
⸱⸱⸱⸱⸱⸱	LCC Becontree Estate boundary
▒	Open countryside/Green Belt

Figure 7.1 Ilford, Dagenham and the Becontree Estate Excursion

most British town centres, the lack of suburban shopping opportunities has resulted in congestion, the need for pedestrianization and the development of off-street shopping malls, inner ring roads and multi-storey car parks about the centre. In Ilford a ring road of sorts was built in the 1980s and access to the main street was restricted. The problem of congestion has been exacerbated by the opening of a superstore in the town centre, at a time when other places were encouraging such high car-generating shopping to go to off-centre sites. Ilford's importance as a regional centre was enhanced by the recent extension of the A406 North Circular Road from the M11 Motorway to Docklands and the future East London crossing of the River Thames.

Ilford expanded northwards in the inter-war period, and Eastern Avenue was built as a make-work project during the Depression to act as a by-pass into London and relieve traffic congestion in the town. The paucity of transport in this part of London resulted in the extension of the Central Line of the London Underground in the 1930s, although the line was not opened until after 1945.

(From Ilford Station or High Road take a no. 145 or 364 bus to Dagenham Heathway)

London's Out-County Estates

In the London of the early 1900s it was realized that redevelopment, the eradication of slum property and overcrowding, and the improvement in services could not be achieved without the reduction of population from the inner-London boroughs. Since the English preferred two-storey rather than apartment living, and since there was no large-scale vacant land left in the LCC area, it was necessary after the 1920s to construct public housing outside the county. The LCC entered into agreements with other counties to buy land, build houses and supervise the transfer of people from inner London.

The out-county estates also reflect a new era in house building and estate design. The influence of the Garden City Movement on both public and private housing was felt after the First World War when the Tudor Walters Report (1918) and the Housing Act (1919) resulted in changes to the urban landscape. Central government subsidies were provided for public housing; housing densities were drastically reduced (to a low of 12 per acre); housing and other amenities were greatly improved; larger front gardens and sometimes very long back gardens were introduced; far greater emphasis was placed on greenery and roadside landscaping (to create the illusion of a move to the countryside); roads were no longer grid-iron and the cul-de-sac became a favourite; and terrace housing for the working class was replaced by shorter terraces or semi-detached houses.

LCC Becontree Estate

The land amassed for this monster housing estate centred on Becontree (the medieval name for the area) and spread out over three local authorities in Essex (Dagenham, Ilford and Barking) (Figure 7.1). Since over 50 percent of the estate was in the former borough of Dagenham, it has been called by this name rather than Becontree. This is the largest out-county estate ever. Between 1921 and 1935 approximately 120,000 people, mostly East Enders, were moved to Dagenham, to some 27,000 houses, covering 2,700 acres (approx. 4 sq. miles). Whilst it provided people with much improved housing, it has long since become a beacon of what not to do in urban planning.

It is an unending flat plain of two-storey houses; until recently there were no higher buildings and few apartments. For a community of this size there has always been poor amenities. Dagenham's Civic Centre was marginal to the estate; there was no high-order shopping (it already focused on Ilford's town centre), but some 10 or more local shopping parades similar to the one at Dagenham Heathway Station; there were only nine pubs to serve all these people - vast barns of places compared to the intimate 'local' one would have left in the East End. Little consideration was given by the house-builders to employment. The 600 acre Ford Motor Works to the south at Dagenham Dock was not started until the 1930s; at its height it employed 35,000 people, although only 20 percent of male employees on the estate worked there. So many services, for example schools, lagged behind the houses and the people, mostly because the LCC was only responsible for housing and the other services were the responsibility of the Essex County Council or the fledgling local authority in Dagenham.

It was a one-class community. 89 percent were defined as blue collar in 1958, the highest figure for 157 towns over 50,000 people in England and Wales, and since the figure relates to the borough, not the estate, the figure for the latter would have been even higher. There were very close links to older East End communities. Because services were poor, employment negligible at first and people were planted down with few, if any, friends and relatives around them, there was mass daily and weekend commuting to inner London until the 1940s. It was a very young community at first; 44 percent of the population were under 15 in 1931 and only 5 percent over the age of 50. Daughters and their children were very isolated from their mothers and extended family, and by day even from their husbands.

Although the estate was built for Londoners, nearly a third of the residents came from other parts of Britain. This was because the LCC built houses in excess of demand. Many Londoners living in desperately inadequate housing could not afford the high rents in Dagenham (triple perhaps what they were then paying) or the Underground fares to reach their workplaces in inner London. Also, the estate seemed so remote from the life they knew, and considerable walking was involved in reaching public transport. At first, many families who arrived at Dagenham could not stand the remoteness or the expense and returned to their older communities.

As time has gone on the instabilities have removed themselves: services and employment have improved; the population has aged and largely replaced itself, giving a more balanced age structure; and it is far less working-class as people have bought their rented houses and employment has changed. Dagenham, in 1951, was the quintessential Labour constituency in Britain with the MP having a majority in excess of 30,000 votes; 30 years later, it had become a marginal seat with the Labour majority closer to 2,000.

Whilst Dagenham has matured, it still contrasts quite markedly with the more middle-class areas of Ilford (now part of the London Borough of Redbridge) to the north or Upminster, Romford and Hornchurch (now the London Borough of Havering) to the east; and it still suffers from a down-market reputation. Much was learned from Dagenham, and in the building of the New Towns after 1945 many mistakes were avoided, although it is interesting to note later those that were repeated. The LCC, later the GLC, were to continue with out-county estates after the Second World War, the most notable being at Harold Hill on the edge of Romford.

(From Dagenham Heathway Station turn left (northwards), right on Reede Road, left on Sterry Road, left on Pettits Road, cross Heathway, left on Singleton Road, left on Parsloes Avenue, right on Heathway to Dagenham Heathway Station)

The form and extent of the Becontree Estate can be readily identified. Compared to the pre-1920 areas in Barking and Ilford to the west (and even the inter-war private estates to the east), one can appreciate the street plan being the work of one housing department at one point in time with draftsmen armed with ruler and compass. Grid-iron roads are very clearly out; in fact, very little is parallel to anything else. If there is a pattern (and that is questionable), it is a set of spine roads, in an inverted Y from Chadwell Heath Station and going either side of Parsloes Park, crossed or framed by other main roads; in

Plate 7.1 Housing on the Becontree Estate. Note how the new
homeowner has changed the timber facing of the semi-detached
house (left), and how narrow roads and lack of parking are dealt
with (Photo: H.J. Gayler)

between are secondary and tertiary through roads built to varying
geometric patterns and finally a large number of cul-de-sacs.

Our short walk from Dagenham Heathway Station will show us
many of the features of this housing estate (and the great contrast with
the housing of inner London where many of the original inhabitants
would have come from). These features include the sense of
spaciousness and lower density housing resulting from quite sizable
gardens, shorter terraces of houses and larger lot widths. There are also
the broken sight lines resulting from the curvilinear pattern, varying
house designs and building materials and the greater use of pedestrian
spaces, particularly in the cul-de-sacs. Vernacular architecture,
especially the timbered Essex country cottage, can be seen (Plate 7.1).
Non-residential land uses are clearly confined to certain areas, for
example adjacent to the railway.

When the estate was built, the car was beyond the means of the large
majority of people, and no accommodation was made for it. This is no
longer the case, although car ownership is still lower here than in some
adjacent areas. Likewise, the parade of shops adjacent to the station is

less than adequate for today's needs. This is, in part, solved by the off-street shopping mall; but the price-conscious or car-oriented shopper, geared to higher-order or superstore purchases, is forced to go to Ilford or Romford or one of the many freestanding superstore locations off the estate.

In the last 20 years there has been a concerted effort to sell public housing to sitting tenants and on the open market. Whilst homeownership has social and economic advantages to both former tenant and taxpayer, the lack of public housing replacement has frustrated housing shortages. Meanwhile, many new homeowners have not survived recession and unemployment, and repossession of homes has been common in places like Dagenham. In good British class fashion a homeowner takes little time to indicate he or she is no longer a tenant: a new and individual front door is usually the first change. Various larger changes, especially structural ones, can be seen, indicating a need or desire to upgrade the 60 year-old housing.

(From Dagenham Heathway take the District Line of the London Underground to Upminster, change to a British Rail train to Basildon)

After 1945 the election of a Labour government and the much greater attention that was to be paid to land-use planning resulted in a very different urban landscape. London was surrounded by a Green Belt which separated the inter-war suburbs from the towns and villages in the Green Belt itself and beyond. London was to change in one of two ways: massive slum clearance and redevelopment, which we have already seen in the 19th-century city (see Chapter 6 above); and the development of New Towns and the expansion of existing towns beyond the Green Belt.

Our excursion today will continue by looking at various aspects of London's expansion beyond the Green Belt and the present County of Greater London. It will focus on one of London's New Towns in the southern part of the county of Essex. Essex is one of London's Home Counties which has experienced a phenomenal rise in population since the Second World War.

London's Green Belt

The end of the District Line of the London Underground was essentially the end of inter-war London, and after Upminster Station we pass into the Green Belt. It is important to remember that the Green Belt is by no means an entirely green area. There were already sizable

Plate 7.2 London's Green Belt, showing the wedges of open space and the 'fossilized' areas of housing that remained once the measure was adopted (Photo: H.J. Gayler)

communities that had developed earlier along the major rail lines out of London and were included in the belt. This would preclude much of their own development in the post-war era, forcing development to go beyond the Green Belt.

Because of this existing urban development in the Green Belt, the actual green area can take on some peculiar shapes (Plate 7.2). Rather than a belt, it can be a field here, a field there. One wonders at times is there any point in retaining these odd parcels of land as fields, or what passes for a field, when they are nearly encircled by urban developments and perhaps crossed by major arterial roads. (This occurs, for example between Dagenham and Havering.) For the most part these 'fields' have been vigorously retained, not so much for their agricultural value but as a buffer between urban communities and to give people a sense of countryside.

It was for reasons of preserving the Green Belt that many development proposals have failed over the years. In South Essex the private enterprise New Town at Tillingham Hall (to the south of West Hordon Station) was rejected during the 1980s, at a time when demand-led planning and private initiative were being greatly encouraged by the

Thatcher government. The Green Belt was a windfall in that it allowed for the relatively easy positioning of the M25 Motorway around London; it was first mooted in Abercrombie's Greater London Plan in 1944 but took another 40 years to come to fruition (Figure 7.2).

Britain's New Towns

The idea for planned New Towns dates back to the Garden City Movement of the late 19th century. Letchworth and Welwyn, and possibly Manchester's overspill estate at Wythenshawe, were the only examples in England; and although the inter-war years were to see a very different kind of development (whether public or private), the idea was kept alive by the Town and Country Planning Association and recommended in Abercrombie's Greater London Plan (Figure 7.2).

The election of a Labour government in 1945 resulted in the appointment of a committee to study the issue, and very quickly it made the recommendation that New Towns should be built to overcome various population and economic problems. They should vary in size from 20,000 to 60,000 people, be self-contained communities, financed by the central government and developed and administered by a local public corporation appointed by central government. This was a departure from the Garden City Movement's idea of a locally-based association of the residents in the town. But at least this top-down financing and planning meant that the New Towns would be a success.

The New Towns Act was passed in the summer of 1946 and the first of the London New Towns was designated in the November of that year. Between 1946 and 1950 eight towns were designated around London, with a further two in Scotland, two in north-east England, one in the Midlands and one in Wales. (The non-London New Towns were built to solve area economic, rather than population, problems.) Only two of Abercrombie's eight New Towns, Stevenage and Harlow, were endorsed by the Government; the others were rejected on the basis of local objections or the loss of valuable agricultural land. Six new sites were chosen, and some of these were not exactly popular. Welwyn was chosen, perhaps because it had hardly got off the ground as a Garden City. We will be looking at Basildon, which was chosen because the area had become a vast rural slum.

After the Conservatives came to power in 1951, there was a major policy swing from the public to the private sector in most areas of the economy. More private than public housing was built, and apart from one more Scottish New Town there were no further designations until the 1960s.

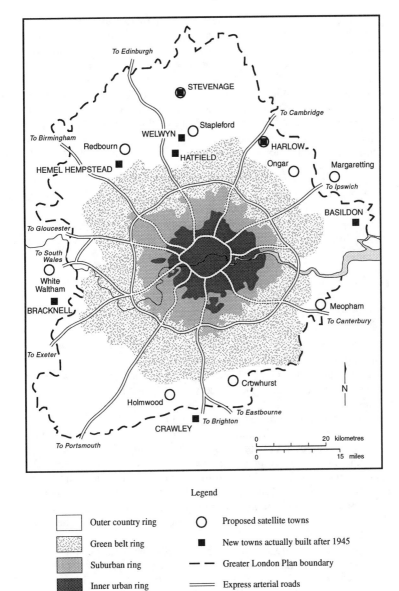

Legend

☐	Outer country ring	○ Proposed satellite towns
▨	Green belt ring	■ New towns actually built after 1945
▨	Suburban ring	– – Greater London Plan boundary
▮	Inner urban ring	═══ Express arterial roads
		········· Arterial roads

Figure 7.2 Abercrombie's Greater London Plan, 1944
(after Hall, 1989)

After 1960 two problems were coming to the fore which were answered with the designation of more New Towns and the expansion of some of the existing New Towns beyond the 60,000 mark. The problems were 1) the continuing high birth rate after the war and the migration of people into southern England which made a joke of the population sums produced by Abercrombie and others at the end of the war; New Towns, especially in the London area, would soak up this increase in people and employment; 2) the decline in economic fortunes of the older industrial areas of Britain in the late 1950s and the need to rehouse people from older inner city areas. Between 1960 and 1970 two more towns were designated in Scotland, one more in north-east England, four in north-west England (Lancashire-Cheshire), one more in Wales, two more in the Midlands and three massive New Towns to service London. The sheer size of the new designations, the downturn in the birth rate and greater emphasis since the 1970s on private rather than public enterprise have resulted in no further New Towns being created.

Basildon New Town

The development of south-east Essex

This area has a colourful history. The New Town was a compromise site after one nearby was rejected. Essex County Council asked the government if it would designate the area a New Town because it was rural slum, lacking services, a health risk and beyond the means of the county to clear up the mess in a short time. The background to this slum developing were the agricultural depression of the late 19th century, the heavy clay soils of the area and the higher than average amount of land going out of cultivation. The keen rivalry between the two railway companies serving south-east Essex (the London, Tilbury & Southend from Fenchurch Street and the Great Eastern from Liverpool Street) had resulted in fairly cheap fares and the development of Southend as a dormitory suburb and a seaside resort. A third aspect to railway business was the development of the 'plot lands'.

Along the two railway lines after the 1890s, farmers, speculators and land development companies carved out roads (in grid-iron fashion), legally divided the land into small plots (usually 18-25 ft. wide), advertised land sales in London and ran trains to the nearest station on the day of the sale. It was hoped to attract weekend cottagers, retired people and commuters. No thought was given to services; it was expected that they would be provided by local councils or private utility companies. Although there were minimum prices set for land sales and the house which could be built, the enormous surplus of land and the lack of planning controls until the 1940s meant that relative location would be the best guide (other than the farmer not selling his land) as to what kind of urban development took place.

In the Southend suburbs, especially close to the seafront, quality homes were built: in the back 'estates' in Laindon and Pitsea, amid the scrub, were hundreds of substandard shacks, built by their owners who would cart the building materials by car or train on the weekends. Most were little better than a garden shed or garage, and there were many railway coaches and bus bodies to add to the colour. Services were a tap at the end of the street (if that), perhaps a gas line, an outhouse with collection service, the village shops, school and pub, and unmade roads that became impassable after a storm. This rural utopia had a boy-scout/girl-guide camp feel to it, and for many it was a paradise compared to London's East End where they may have lived.

The inter-war period saw more extensive 'plot land' development around London and the permanent occupation of many of the homes. The blitz of 1940-1941 and the critical housing shortage as troops returned from the war resulted in the meanest shack being occupied. Between 1945-47 the population of the Laindon-Pitsea area increased threefold to 20,000. There were fears of an epidemic of massive proportions.

The post-1949 'New Town'

Unlike the other London New Towns, Basildon (taking its name from a small hamlet between Laindon and Pitsea) had nearly 25,000 people already living in its designated area of 7,800 acres; approximately 4,500 of these acres consisted of 5,600 substandard shacks along 78 miles of unmade road. In the designated area there were 30,000 different land and building ownerships (thousands of 'plots' were purchased but never built upon). Many people, or their descendants, did not even know they owned a plot of land at Basildon; between 1949 and 1966 28% of the 4,790 acres of land purchased by the Basildon Development Corporation were from unknown owners.

A Master Plan was drawn up for the designated area which would result in the almost complete eradication of the shacks. The basic unit of the Plan was a neighbourhood, and as each new neighbourhood was built in a east to west progression, people would be moved in groups from the next area of shacks to be demolished. This allowed for the continuation of communities in a way that was not present in the LCC overspill estates or urban renewal programmes (Plate 7.3).

Plate 7.3   Basildon, looking north from the town centre, showing a
residential neighbourhood, a large area of parkland and an industrial
park in the background (right) (Photo: H.J. Gayler)

The joining of the two 'villages' of Laindon and Pitsea required a
larger population than the traditional 60,000 maximum.   It was
subsequently increased to 140,000, the approximate figures chosen for
the 1960s London New Towns of Milton Keynes, Northampton and
Peterborough, and reflected the greater opportunities available with a
larger population base and the less likelihood of commuting out for
jobs and recreation.

*(From Basildon Station cross Southern Hay into Podderwick,
turn right on Town Square)*

The focus of the New Town was a new pedestrianized town centre
just north of the Fenchurch Street-Southend rail line, half way between
Laindon and Pitsea near the hamlet of Basildon (Figure 7.3).   This was
completed in the early 1960s and was an open mall of the type found
anywhere in renewed British town centres.   Twenty years later
continued population growth in the designated area and outside resulted
in the addition of a large superstore and enclosed multi-level shopping
centre.   Basildon is now an important regional shopping centre between

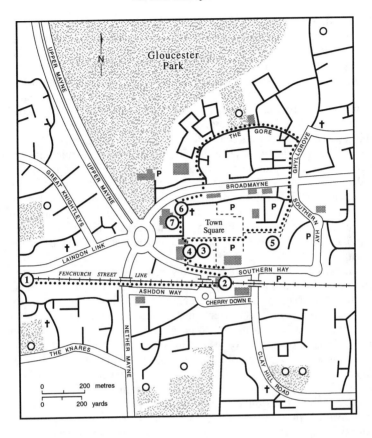

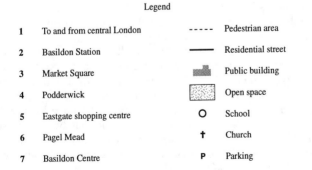

Legend

1	To and from central London	- - - - -	Pedestrian area
2	Basildon Station	——	Residential street
3	Market Square	▨	Public building
4	Podderwick	▨	Open space
5	Eastgate shopping centre	O	School
6	Pagel Mead	†	Church
7	Basildon Centre	P	Parking

Figure 7.3 Central Basildon Walk

Romford, Southend and Chelmsford. Laindon and Pitsea, along with
the northern communities of Billericay and Wickford (just outside the
designated area), are district centres for Basildon.

The railway station in the town centre was a late addition, after
years of arguing that to provide one would encourage commuting to
London. An essential feature of the Master Plan was to provide a large
industrial area between the town centre and the Southend Arterial Road
to the north. The growth of industry was, however, slow because until
the 1970s, on paper at least, there was a government policy to
encourage industry to develop in those parts of Britain that suffered
from high unemployment and a concentration of old declining
industries. It was important that the town be self-contained, although
as always retail services, offices and recreational facilities lagged
behind housing development. However, there was less turnover of
people compared to places like Dagenham because a high proportion of
the planned migration from the London area was tied to jobs in the
town.

As was the case in all New Towns, the government-appointed
Development Corporation only built homes for rent at first. This
produced a largely working-class community with services to match.
Around the New Town were more middle-class (and London
commuter) towns, and these people would commute to Basildon to
work. Later, more homes were built for sale and in the 1980s there was
a government-sponsored policy to sell public housing at below market
value to sitting tenants. This has altered the social composition of the
town; and while the very word Basildon lingers in the British psyche as
a large public housing estate, the Socialist tenant has now been
transformed into 'Essex man' (or 'Essex woman'), the media-
portrayed, upwardly-mobile, Conservative-voting homeowner of new-
found but modest wealth.

Once the New Town was completed the Development Corporation
was wound up and its remaining operations handed over to the local
council. Today, there is no difference administratively between the
designated New Town and the rest of the town (once named Billericay
after the largest community and renamed Basildon). The fact that there
was a Master Plan and central government financial backing meant that
Basildon was no spartan LCC housing estate. Also, the Master Plan,
now part of the Local Plan for all of Basildon, has been updated to
account for urban change; for example, the impact of greater car
ownership, the rise of tertiary and quaternary employment, the switch
from neighbourhood shopping to superstore purchasing, the growth of a
socially more balanced community and the emergence of Basildon as a
regional centre for south Essex.

(From Town Square pass Eastgate Centre, turn left at Southern Hay, cross Broadmayne into Ghyllgrove, left on the Gore, right on Broadmayne, left on Pagel Mead, ahead to Podderwick, cross Southern Hay to Basildon Station)

Our walk in the central area of Basildon (Figure 7.3) will identify some of the important features of the New Town discussed above:-

1)   *Roads*: a three-level system of trunk routes with little or no direct access to homes or businesses (e.g. Southern Hay), secondary through routes within residential areas (e.g. The Gore) and residential streets and cul-de-sacs often connected to other roads by pathways.

2)   *Trunk routes*: these surround the town centre featuring access to all important services (bus, train, police, fire, ambulance, library, council offices and car parks).

3)   *Town centre*: this is fully pedestrianized with an older shopping area around a Town Square, an open-air market (in the style of the street markets of the East End) and a multi-level enclosed shopping mall (the Eastgate Centre).

4)   *Superstore*: this facility serves a wide area of the town and beyond. No doubt its trunk-route accessibility encouraged its location and retention here, when elsewhere supermarket-style shopping has been leaving town centres.

5)   *Ghyllgrove*: this typifies the original concept of a self-contained neighbourhood with its school, community centre, church, local shops, public open spaces, walkways and cycle paths that connect residential areas and road access that restricts through traffic.

(Return to central London from Basildon Station to Fenchurch Street; transfer to Tower Hill Underground Station)

Further reading

Gayler, H.J. 1970.   Land Speculation and Urban Development: Contrasts in South-East Essex, 1880-1940. *Urban Studies* 7: 21-36.

Hall, P. 1988. *Cities of Tomorrow*. Oxford: Basil Blackwell, 86-135.

Hall, P. 1992. *Urban and Regional Planning*. 3rd. edition. London: Routledge, 90-158.

Rubinstein, A. ed. 1991. *Just Like the Country: memories of London families who settled the new cottage estates, 1919-1939.* London: Age Exchange.

Willmott, P. 1963. *The Evolution of a Community: a study of Dagenham after 40 years.* London: Routledge & Kegan Paul.

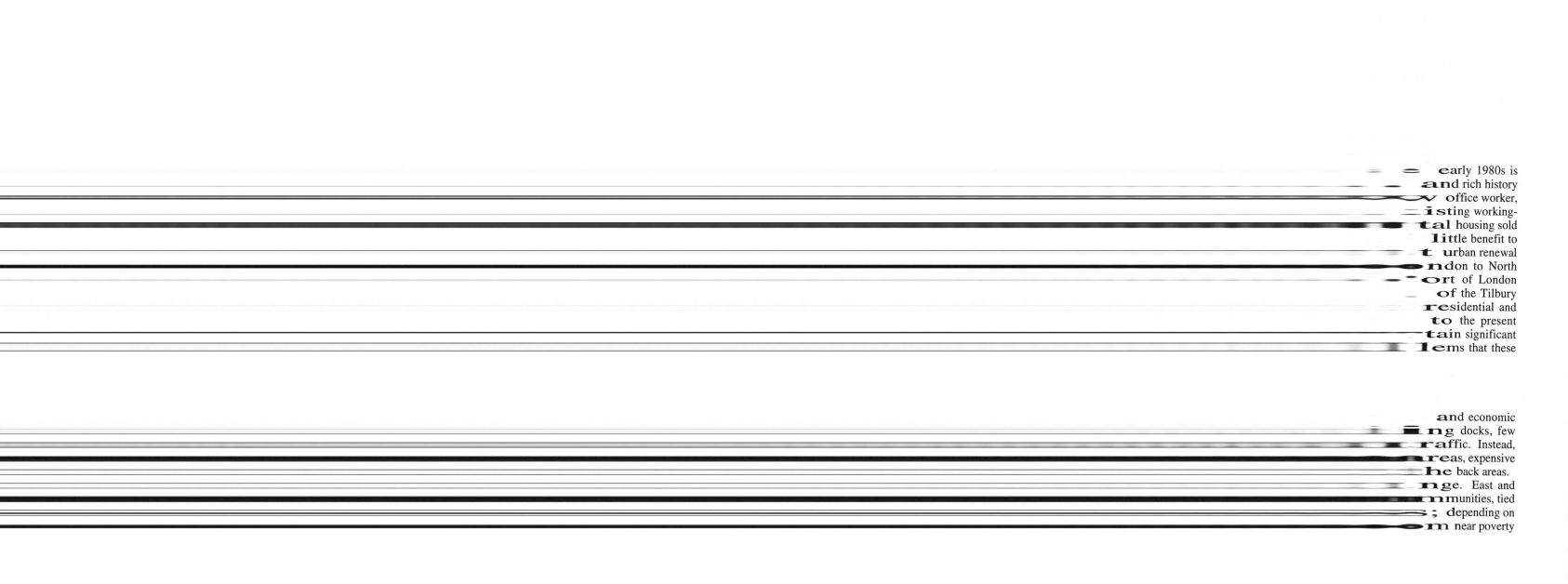

line to desperately poor.

example, were not regularly

were ships to load or unlo

Victorian slum - a brutal,

housing, high death ra

employment), poor servic

century there had been a n

working class, especially n

net effect had not been to

1940-1941 was to have a

measures were to encourag

example, the London Coun

after 1920.

Dockland population ha

century, but the bombin

dispersal of population

approximately one third

government at Westminste

State and vast sums of pu

other services. We exami

suffice it to say that the

materialized. Many of the

95 percent of the housing

and even some of the nev

Then the employment bas

the mid-1960s. 18,000 do

1981, and local unemploy

In the past there has o

certain services and proba

that other Londoners rarel

lived close to their empl

communications between th

Before 1980 only a few m

since 1980 there has bee

housing (at first to the sitt

and much of the private h

population of the area feel

traditional working-class

reach have further antagoni

Along with social cha

1920s most of these Dock

MPs in the House of Con

majorities, backed by a v

(From Town Square pass Eastgate Centre, turn left at Southern Hay, cross Broadmayne into Ghyllgrove, left on the Gore, right on Broadmayne, left on Pagel Mead, ahead to Podderwick, cross Southern Hay to Basildon Station)

Our walk in the central area of Basildon (Figure 7.3) will identify some of the important features of the New Town discussed above:-

1) *Roads*: a three-level system of trunk routes with little or no direct access to homes or businesses (e.g. Southern Hay), secondary through routes within residential areas (e.g. The Gore) and residential streets and cul-de-sacs often connected to other roads by pathways.

2) *Trunk routes*: these surround the town centre featuring access to all important services (bus, train, police, fire, ambulance, library, council offices and car parks).

3) *Town centre*: this is fully pedestrianized with an older shopping area around a Town Square, an open-air market (in the style of the street markets of the East End) and a multi-level enclosed shopping mall (the Eastgate Centre).

4) *Superstore*: this facility serves a wide area of the town and beyond. No doubt its trunk-route accessibility encouraged its location and retention here, when elsewhere supermarket-style shopping has been leaving town centres.

5) *Ghyllgrove*: this typifies the original concept of a self-contained neighbourhood with its school, community centre, church, local shops, public open spaces, walkways and cycle paths that connect residential areas and road access that restricts through traffic.

(Return to central London from Basildon Station to Fenchurch Street; transfer to Tower Hill Underground Station)

Further reading

Gayler, H.J. 1970. Land Speculation and Urban Development: Contrasts in South-East Essex, 1880-1940. *Urban Studies* 7: 21-36.

Hall, P. 1988. *Cities of Tomorrow*. Oxford: Basil Blackwell, 86-135.

Hall, P. 1992. *Urban and Regional Planning*. 3rd. edition. London: Routledge, 90-158.

Rubinstein, A. ed. 1991. *Just Like the Country: memories of London families who settled the new cottage estates, 1919-1939.* London: Age Exchange.

Willmott, P. 1963. *The Evolution of a Community: a study of Dagenham after 40 years.* London: Routledge & Kegan Paul.

Chapter 8

London's Docklands

The redevelopment of London's Docklands since the early 1980s is highly controversial. Much of the earlier development and rich history are being eradicated, converted or gussied up for the new office worker, the tourist and the yuppie resident. Meanwhile, the existing working-class population of the area has seen jobs disappear, rental housing sold off and new services provided which are frequently of little benefit to them. We will be examining what is the world's largest urban renewal project, extending downstream from the Pool of London to North Woolwich and Beckton and including all of the Port of London Authority's dock and wharf areas (with the exception of the Tilbury Docks, 25 miles from central London) and adjacent residential and industrial land. We will focus on the background to the present changes, the organization undertaking the renewal, certain significant developments in the area and some of the major problems that these developments are generating.

Economic and social change

This urban renewal is resulting in a mass physical and economic transformation of the area. There are now no working docks, few warehouses and industries, and little commercial river traffic. Instead, there are office blocks, converted warehouses, shopping areas, expensive new condos, yacht basins and more affordable houses in the back areas.

Concomitant with this has been a mass social change. East and south-east London had long been solid working-class communities, tied either to manufacturing industry or dock and allied trades; depending on the state of these industries and trades, families varied from near poverty line to desperately poor. Until the post-war period, dock workers, for example, were not regularly employed, but were called if and when there were ships to load or unload. In many ways this area was the classic Victorian slum - a brutal, overpopulated and congested area of bad housing, high death rates, high unemployment (or irregular employment), poor services and poor prospects. Since the late 19th century there had been a number of measures to improve the lot of the working class, especially new housing and improved services, but the net effect had not been to change the area significantly: the blitz of 1940-1941 was to have a greater effect. Also, many of the earlier measures were to encourage the better-off families to leave the area; for example, the London County Council's out-county estate developments after 1920.

Dockland population had been declining since the earlier part of this century, but the bombing, evacuation and planned and unplanned dispersal of population resulted in the 1981 population being approximately one third that of 1931. The election of a Labour government at Westminster in 1945 heralded a new era of the Welfare State and vast sums of public money have been spent on housing and other services. We examined these in detail in Chapter 6 above, but suffice it to say that the utopia planned for the East End never materialized. Many of the improvements never stood the test of time. 95 percent of the housing was rented, mostly from the local council, and even some of the new housing was in a deteriorating condition. Then the employment base for the East End completely fell apart after the mid-1960s. 18,000 dock-related jobs were lost between 1966 and 1981, and local unemployment rose to 24 percent by 1981.

In the past there has only been a small middle-class, employed in certain services and probably commuting in each day. It was an area that other Londoners rarely visited; together with the fact that workers lived close to their employment resulted in there being very poor communications between the Dockland areas and other parts of London. Before 1980 only a few middle-class trend setters elected to live here; since 1980 there has been a veritable flood. With the sale of public housing (at first to the sitting tenant), little or no new public housing, and much of the private housing beyond their reach, the working-class population of the area feels especially squeezed. Moreover, the loss of traditional working-class jobs and most of the new jobs beyond their reach have further antagonized the original population.

Along with social change has come a political change. Since the 1920s most of these Dockland areas have been represented by Labour MPs in the House of Commons and local councils with large Labour majorities, backed by a very strong trade-union movement. At times local councils have had 100 percent Labour membership. In the last decade this has been changing as Liberal-Democrat MPs and local councils have been elected, and various centre-right groups, such as the British National Party, have exploited racial and economic problems to attract the White vote.

In true London fashion the extensive changes in Docklands are still an example in planning of incrementalism rather than some grand design; a case of muddle through and hope the blunders will not be too obvious rather than a rational, comprehensive, visionary and well-financed plan from the beginning. Our excursion will look at the past, present and future of this monumental exercise in laissez-faire, demand-led planning.

The Development of the London Docks

The development of London as a port city is closely associated with the growth of Britain as a major colonial and industrial power. Until the 1960s it was the largest port in the world, assuming that role in the medieval period. It was well placed to serve England when nearly all trade was with continental Europe. And in spite of the fact that West Coast ports were better placed to serve the New World (for example, Bristol and later Liverpool, Manchester and Glasgow), London had the advantage of inertia, a market far larger than any other British city, a cost and time disadvantage for freight that was not excessive and the fact that so many merchant companies were London-based and could grant charters ensuring a monopoly of all English trade through the Port of London.

The growth of the port meant congestion alongside the river wharves. Although small wet docks had been constructed in the late 1600s, it was not until 1799 that a Royal Commission reported in favour of wet docks and the first Dock Act was passed. The West India Docks on the Isle of Dogs were opened in 1802, followed by the London Docks in 1805, East India Docks in 1806 and St. Katherine's Dock in 1828. The Surrey Commercial Docks, the only dock system on the south bank, were opened during the 19th century, as were Victoria (1855), Millwall (1868), Royal Albert (1880), Tilbury (1886) and King George V (1921). Severe financial difficulties and lack of important maintenance resulted in the companies coming under public ownership (the Port of London Authority) in 1908.

A significant proportion of the incoming traffic to the docks was either transferred to lighter or barge for transportation upstream to riverside wharves (coal for power stations, for example), or went out of the port by rail or coastal steamer. Road access to the port was poor;

also the docks were surrounded by high walls and few gates to reduce pilfering. The post-war increase in road traffic to the docks began to cause congestion. Furthermore, apart from Tilbury, London was poorly placed for containerization; there simply was not enough room to handle the goods. Also, London had to face new mechanical handling methods and the rising cost of unionized labour. Together with the decasualization of dock labour and a general decline in industrial trade, the dock trades went into a complete tail-spin in the 1967-70 period. Increasingly, other ports, for example, Felixstowe and Hull on the East Coast or Rotterdam in Holland, could do a better job. Beginning with the East India Docks in 1967, all of London's docks except Tilbury were to close within the next 15 years.

The Way Forward for Docklands after 1970

The collapse of the dock trades, massive industrial closures and the continual decline in population quickly transformed the East End into one of Britain's most deprived inner-city areas, attracting the momentary attention of the Prince of Wales and the publicity he brings. In the 1970s this growing area of dereliction caused concern amongst the various public and private bodies, including the now-defunct Greater London Council (GLC), the local London boroughs (Tower Hamlets, Newham, Southwark, Lewisham and Greenwich), the Westminster government, the Port of London Authority, trade unions, and residents' groups. In 1974 the GLC established a Docklands Joint Committee, comprising representatives of the above organizations, and in 1976 a London Docklands Strategic Plan was published. Planning philosophy focused on the needs of local areas with financing mainly from the public sector. However, little was done about the problem; the area lacked a single, legal authority with the necessary powers to co-ordinate redevelopment and the various London boroughs competed rather than co-operated.

The election of a Conservative government in 1979 brought about fundamental changes in administration, planning philosophy and development initiatives. What we see today are the results of this new initiative.

London Docklands Development Corporation (LDDC)

The key to change was a new approach to administration. In 1981 the Thatcher government established the LDDC with the statutory powers within a certain area (see Figure 8.1) to plan, buy and sell land,

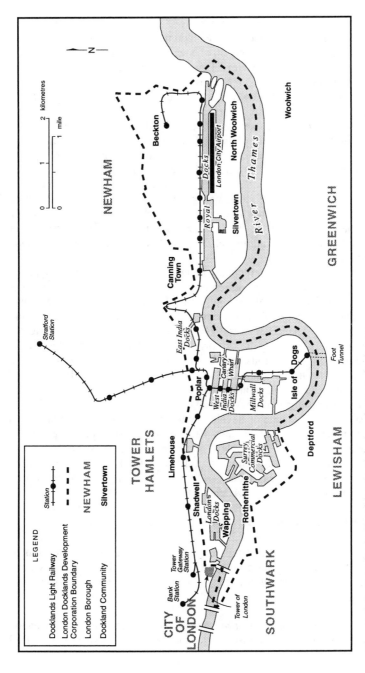

Figure 8.1 London's Docklands showing former dock areas

improve infrastructure, and promote development by the public and private sector, i.e. to facilitate development rather than be a developer. The LDDC was modelled on the various New Town Corporations formed after the 1946 New Towns Act (except those corporations actually built houses, factories, shops etc.).   It was a central government appointed body, responsible to the Treasury for its finances and divorced from local governments and purely local needs.   The two-tier system of local government (County and borough) found itself with part of its area where its authority under various Acts of Parliament (for example, the Town and Country Planning Act) no longer applied. Moreover, part of the LDDC area in the Isle of Dogs was designated an Enterprise Zone in 1980, further removing an area from local government control.   Central government could thus impose its will on local areas whether they liked it or not.   Given the conflict between a Conservative central government and Labour local administrations, as well as opposing planning philosophies, the battle lines were soon drawn.

The basis for development in Docklands was to be demand-led planning, as opposed to structure planning on the basis of some pre-determined set of goals.   In a way it was back to the incrementalism that government had been trying to get away from since the Second World War.   The LDDC was to be concerned with wider market-oriented demands based largely on private investment.   This would be encouraged by various tax breaks and cutting back severely on traditional bureaucratic controls, for example the long delays that a developer goes through under the Town and Country Planning Act.

A major element in the LDDC's mandate was to introduce a completely new type of development into the Docklands area, aimed very much at the middle-class resident, visitor and employee and the hope that these new peoples, investments and jobs would have a beneficial 'trickle-down' effect on the local working-class population. Three types of development can be seen: infrastructure improvement, employment change and residential development.

Infrastructure improvement

Both the existing population and potential new development demanded improved service provision.   The following areas have received attention, although in almost all cases need and provision hardly coincide:

1)   *Public Transportation*   For an area so close to central London Docklands never had a fast direct public transport service.   British Rail

connections between Liverpool Street or Fenchurch Street and Docklands had ceased long ago, reflecting a very locally-based population in terms of employment, shopping, and leisure activities. Similarly, on the south bank the London Bridge lines were ill-placed to serve the area, and there were only stations at Deptford and Greenwich that were anywhere close. The London Underground had one antiquated line from Whitechapel Road on the Metropolitan/District lines, serving part of the London and Surrey Commercial Docks en route to New Cross or New Cross Gate. British Rail had a cross town line from the north-eastern suburbs to North Woolwich, serving the most easterly Royal docks. Bus connections to anywhere in London were slow and often circuitous in order to maximize passenger loads.

Part of the LDDC's strategy was to improve transport. The most significant change has been the opening of the Docklands Light Railway (DLR) in 1987, largely on the alignment of defunct British Rail tracks (and on a viaduct which promotes good views!). It is run by the LDDC and is integrated with the Underground and British Rail systems. But unlike the rest of the system it is a light rapid transit type rather than the traditional heavy-rail, and does not physically connect with the other lines. Until the Bank extension was opened in 1991 the line terminated at Tower Gateway, a separate station from either Fenchurch Street or Tower Hill nearby. A similar situation exists as Shadwell and Bow Road/Bow Church. Only at Stratford is there a direct connection with British Rail and the Central Line, although the Beckton extension will also have a British Rail connection.

Very quickly this Toytown-Disneyland style railway was carrying loads far in excess of those planned, including local residents and workers, tourists exploring on their travel cards, school parties on their way to Greenwich etc. There was scant regard for (or no knowledge of) the heavy use that could be expected from office workers, especially the massive developments proposed for Canary Wharf. The DLR has now been extended to Beckton (where it largely serves the derelict Royal dock area!), and there are plans to go south of the Thames from Island Gardens to Lewisham. It does not solve the load and access problems to and from central London. The expansion of the trains and platforms from two cars to four will no doubt help, but the key feature will be the extension of the Jubilee Line from Green Park to Waterloo, London Bridge, Surrey Quays, Canary Wharf, Port Greenwich and Canning Town to Stratford.

Bus transportation has been improved but it still essentially acts as a feeder service to the DLR and other rail lines; although with the DLR out of action so much of the time for improvements and extensions, bus services have come into their own!

2) *Road improvements* Like rail transportation, this has also been a case of planning too little, too late. Outside the central area the majority of work trips are by private transportation, and Docklands would eventually be no exception. However, the roads could hardly be considered adequate. The A13 Commercial Road-East India Dock-Newham Way is already a major arterial route between London and the north side of the River Thames. It found itself acting as the spine road for Docklands, and it has increasingly been unable to perform this task. New east-west arterial routes have now been built closer to the river to improve the situation. Routes into east and south London via M11, M2/A2 and M25 and their extensions are much improved. However, local and arterial roads in the inner area are often inadequate. Of note are the poor communications across the River Thames, including two old road tunnels with difficult exits and the Woolwich Ferry: the proposed East London crossing connecting major arterial routes from Beckton to Thamesmead and beyond is desperately needed.

3) *Air transport* The long wharf between two of the Royal docks in Silvertown-North Woolwich has become London City Airport. Its proximity to both Docklands development and the City of London was to encourage growth and relieve London's major airports. However, growth has not been that impressive, connections are limited, range is restricted even with the introduction of jet aircraft, and local residents are opposed to the noise aspects and worry about planes being too close to houses. It is interesting to note that the airport terminal does not connect directly with rail transportation, and nor are there any plans to do so!

4) *River transport* A regular, speedy, river-bus service between various piers from Chelsea in west London to Greenwich for commuters and tourists was tried for a number of years. But financial difficulties, resulting in part from insufficient traffic, have forced the system to shut down. There are still a number of summer pleasure boat services between central London and Greenwich.

5) *Shopping facilities* Prior to the Docklands redevelopment the shopping facilities of the area were sadly lacking. They were local in nature, reflecting the myriad local communities and low levels of car ownership, and were strung out along busy main roads. The advent of the large, modern supermarket (let alone superstore) seemed to pass

much of east and south-east London by until recently. Higher-order shopping was often distant to these communities, and the West End was not part of these peoples' mental map.

However, it is to be wondered whether the LDDC has improved the situation. There was no retail hierarchy planned, as would be normal elsewhere. Instead, there are trendy shops for tourists, the middle-class and the office workers at Hays Galleria, Canary Wharf and Tobacco Dock; and there are new district (community) centres, featuring large superstores, retail warehouses and smaller stores on the Isle of Dogs and at Beckton and Surrey Quays. The same sad collections of local stores are still to be seen in various older communities (for example, Cubitt Town), and regional shopping facilities are still as distant.

6) *Leisure facilities* Again, something that was lacking before redevelopment, but now there is a much greater emphasis on private rather than public investment; but it has to be questioned whether the facilities that have been built are for the population at large or the middle and upper-income sector of it. For example, the conversion of the docks into yacht basins and areas for water sports, or the construction of the Beckton Alps simulated ski hill. There is little doubt that the provision of new parks, leisure centres, urban farms and a range of different museums and cultural events is aimed at the wider population, but user fees frequently inhibit their use.

Employment change

The infrastructure improvements were designed to support an increase and a major change in direction for employment in Docklands. For a long time it had been thought that the City of London and parts of the West End could not take more office employment. Unless London was to see more high-rise developments akin to North American cities, and even more transportation improvements to get the commuters in, then some alternative would have to be found. For years there had been various restrictions on office expansions in the central area (and the number of commuters has declined slightly), encouraging developers to seek more suburban or provincial town locations. Certainly, the high rents in London were an incentive to move out. Docklands was manna from heaven - a non-central site, where rents would be cheaper, but close enough to the City of London to be considered an extension of it.

One of the early exports to Docklands was the printing and publishing industry from Fleet Street, including the News International papers to Wapping (known as Fort Wapping because of the lock-outs, strikes and police action) and the Guardian and Telegraph newspapers to the Isle of Dogs. Other smaller industries also came and small shed-like buildings were constructed, especially in the Enterprise Zone on the Isle of Dogs. The more recent boom in office development, along with recession and rationalization, has resulted in redevelopment already, and some of the sheds have given way to more impressive buildings.

Little consideration was given as to whether the new jobs could be taken up by the existing population. As the dock trades and associated industries wound down, unemployment shot up to nearly 25 percent in some places. However, unemployment persists at a high level because the skills of so many people do not match what is needed without considerable retraining and further education. The 'trickle-down' effect is often illusory.

The major impetus in employment change has undoubtedly been the enormous office development at Canary Wharf, undertaken in the late 1980s by the Canadian firm, Olympia & York. This expansion was not foreseen a few years earlier when the modest infrastructure changes were being made by the LDDC. A more detailed explanation will be seen below during the actual excursion.

Residential development

The LDDC area never had a large residential population. So much of the land was taken up by docks, industry and warehousing, and what housing did exist has been subject to a wartime blitz and slum clearance programmes. Some of the Dockland working-class communities had lost so much population, it could be argued that they hardly functioned well. Could the school be kept open? Was there sufficient threshold to attract better shopping facilities?

In the last decade all this has changed in a variety of new developments, in the context of emerging, enlarged and enriched communities, principally Wapping and Poplar, Isle of Dogs, the Royal Docks (which includes Beckton) and Surrey Docks (which includes the riverside areas between there and London Bridge). Within these communities are a number of distinct neighbourhoods, and the various housing developments have been organized with these in mind. We will see the following types of new developments:

Riverside condos: these are usually low-rise apartment developments, and are either purpose-built or conversions of old warehouses. They comprise the most expensive housing in Docklands.

Terrace housing: this is more likely to be found in the dock areas and converted industrial land.

Semi-detached, detached/link homes: these lower density and often more affordable homes are found principally in the back areas - the inner part of the former Surrey Docks and the far-flung reaches of Docklands at Beckton.

This housing is found in a multitude of designs (although most could be called postmodern), arrangements, building materials, colours, sizes, densities, and prices. It has invited architectural and design awards on the one hand, intense criticism on the other (for example, cries of shoddy, boring, inaccessible, poorly serviced, yuppie jungles, and fortresses with their gates and walls). There is little doubt that these new housing developments have brought to inner London an expanding middle-class community that is younger, has fewer children, is better educated, has more income, and is more White than the existing working-class population. The spin-off effects are clear: these people need a whole range of services but it is unclear as to the extent to which the existing population can, or can be trained to, provide these.

Existing housing in the LDDC area has also been transformed in the last 15 years. The majority of this was for rent from the local councils, and it was the policy of the Conservative government since 1979 to force councils, if necessary, sell off most of its public-housing stock to its sitting tenants. For those who were able, this was a financial gain because one could buy low and eventually sell on the normal market for much more. The government also gave financial incentives to tenants and private entrepreneurs to form housing associations to buy and manage many of the apartment blocks that were either in bad condition or had such poor reputations that few would purchase property there.

It was government policy after the Second World War to demolish the worse terrace housing of the 19th century and replace it with medium to high density public-housing apartment complexes. However, some of this new development was so poorly designed and built, as well as poorly maintained, that it is being renovated or demolished less than 30 years later. Until this happens, this housing, which often contains the worse poverty cases in the population, sits in stark contrast to the nearby owner-occupied properties, and thus fertile ground for resentment, vandalism and other criminal activity by an underprivileged group.

The present serious recession in Britain has a deleterious effect on the housing market. Many households encouraged to buy into the property market in the good times can now ill-afford to keep up the mortgage payments on their home in the poor times. There is currently the greatest repossession of homes by mortgage companies that there has ever been. Housing vacancies are high, and property prices in

Britain are still falling. People who have to sell now do so at a loss. This is affecting not just people in former council property but many in the newer and more expensive housing areas of Docklands.

Our tour through Docklands will encompass many of the features noted above (Figures 8.2-8.3). By means of walking, train and bus travel, and a pleasure boat trip back to central London, we can experience this new environment, the remnants of the old, and the clash of culture and class.

(From Bank Underground Station in central London take the Docklands Light Railway (DLR) to Shadwell)

Shadwell

This dockland community grew up first as a ribbon development out from the City of London along Cable Street and the Highway in the 1600s, with infilling between there and the river taking place in the 1700s. The sole magnificent reminder of this past is St. George-in-the-East Church, built by Nicholas Hawksmoor between 1714 and 1730, and one of the many churches built in new suburban areas beyond the Cities of London and Westminster at that time.

Little of the old Shadwell remains. Progressive slum clearance, beginning at the turn of the century, and wartime damage and renewal have all but replaced the two-storey housing by apartment blocks. The Dellow and Bewley developments immediately south of Shadwell Station are among the first apartment blocks built by the LCC after its incorporation in 1889. Many of the other apartments also predate the Second World War, and most are in the process of being refurbished. Local services vary considerably in quality, from a new swimming pool to very inadequate local shops around the two Shadwell stations.

The Highway is the boundary of the LDDC area and certainly a social divide between the old and new developments. Changing road infrastructure in Docklands, especially the new Limehouse link tunnel connecting the Highway to the former West India Docks, has resulted in this road becoming a busy and fast arterial route and a physical barrier to north-south access.

(From Shadwell Station cross Cable Street to Dellow Street, right at the Highway, left on Wapping Lane, right on Pennington Street to Tobacco Dock)

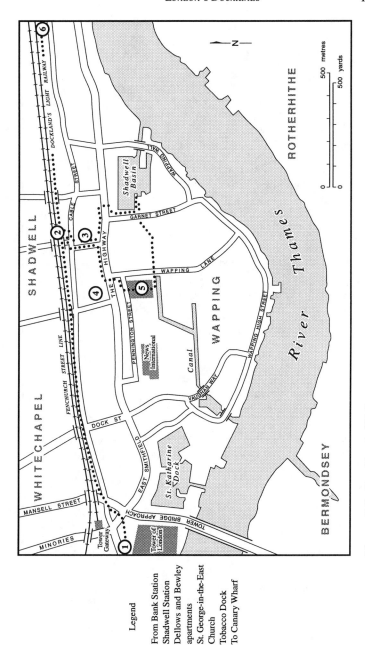

Figure 8.2 Docklands Walk

Legend

1 From Bank Station
2 Shadwell Station
3 Dellows and Bewley apartments
4 St. George-in-the-East Church
5 Tobacco Dock
6 To Canary Wharf

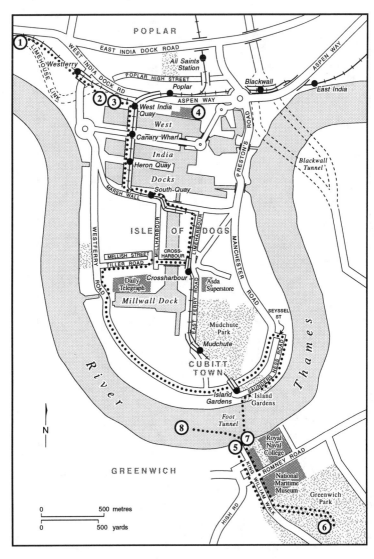

Legend

1	From Shadwell via DLR	5	Cutty Sark
2	Dock Office and Dockmaster's House	6	Old Royal Observatory
3	Gwilt Stacks	7	Greenwich Pier
4	Billingsgate Market	8	To Tower Pier

Figure 8.3 Isle of Dogs - Greenwich Excursion

Tobacco Dock

As the name implies this was originally one of the London Dock warehouses for the importing of tobacco. This was typical of the dock scene prior to the 1970s with formidable high walls to keep people out and to prevent pilfering. Sadly, it is the only dock warehouse to remain. It was converted initially into an up-market shopping centre, dependent on the new yuppie population in the Shadwell-Wapping area, as well as the tourist. However, it has not been a financial success; many stores have remained empty, ownership has changed and new marketing strategies are to be tried. The latest reincarnation is a factory-outlet mall.

To the south of Tobacco Dock lies the filled-in areas of the London Docks, containing middle class condominiums, ornamental canals, old sailing vessels and parkland development.

(Go through Tobacco Dock, turn left on Discovery Walk, follow the canal through to Benson Quay on Shadwell Basin)

Shadwell Basin

This is the only part of the London Docks to remain, and its colourful Legoland-style housing is unique. It is an almost cheeky addition to the urban landscape, and one that continues to draw both praise and criticism. The water body itself looks forlorn and deserted, in common with most of the former dock areas most of the time.

(Left around Shadwell Basin to the Highway, cross the Highway and right on Dellow Street, returning to the Shadwell DLR Station. Take DLR train to Canary Wharf)

West India Docks

The DLR over much of the route from Tower Gateway to Poplar, on the north side of the West India Docks, follows the alignment and the original viaduct of the long-defunct London and Blackwall Railway. As a tourist feature, the DLR affords excellent views of the Docklands developments.

Much criticism has been levelled at the LDDC, the government and developers that so little of the docks' history, especially its buildings, has been retained. En route to Canary Wharf at least something remains of the West India Dock and can be seen on the right, including the Dock office, Dockmaster's house and a former Import Dock warehouse, the Gwilt Stacks. Some of the warehouses were demolished for the new Billingsgate (fish) Market, seen at the east end of the Import Dock.

Canary Wharf

This part of the West India Docks is more than simply another office development in Docklands (Plate 8.1). It is regarded by some as the 'jewel in the crown'; although some would regard it as a major change in the direction of development, of little relevance to the local community, which could possibly lead to massive problems in the future. The arrival in 1987 of Olympia and York, the Canadian firm owned by the Reichmann brothers, changed Docklands from its modest redevelopment program to one that was extensive and rapid.

The project's scale is enormous and audacious:- $8 billion over the following 10 years, 26 buildings (approx. 12.2 million sq. ft. of office space), 250 shops (approx. 500,00 sq. ft.), a 400-bedroom hotel, 60,000 workers eventually on the site, open spaces, cultural events, new roads, rail stations etc. The centrepiece is the 50-storey tower, the highest in Britain, designed by Cesar Pelli, postmodern in design, with an illuminated cap that can be seen for 25 miles in some directions. Canary Wharf's scale was city-centre like, and Docklands was instantly dubbed London's Third City. The design of the project is overwhelming, a throw-back to another era when countries exuded power from their industrial capital and colonial might.

The sheer scale of Canary Wharf has brought problems. It required a considerable increase in the level of infrastructure provision. Suddenly roads, railways, etc. were inadequate, and the LDDC, which was the facilitator of development, found itself with having to provide more things, more quickly, and its spending went into the stratosphere, at a time when the government was desperate to cut back on public spending, lower taxes, sell off public corporations etc.

To add insult to injury, this development was conceived and commenced at a time when a recession was hitting the property market. The price differential between rents in the City of London and the Isle of Dogs was diminishing as new properties (for example, Broadgate) became available in the City and vacancies arose everywhere. Suddenly, the 'remoteness' of Docklands with its growing transportation problems took some of the gloss off the new development. However, the developers would argue that the percentage of space let so far has been impressive, financial woes are exaggerated and transportation problems will be alleviated once the Jubilee Line of the London Underground is completed. However, the financial collapse of Olympia and York has

Plate 8.1 Canary Wharf. This view from the west shows the 50-storey tower, the centrepiece of the former Olympia & York development, and the office blocks at the western end of the complex. On the right is condominium housing on the south bank at Rotherhithe (Photo: Canary Wharf Ltd.)

placed the running of Canary Wharf in the hands of the creditors and there are doubts about if and when the scheme will be completed.

A walkabout on Canary Wharf cannot fail to impress. The immense size and density of buildings and their style imply a powerful presence, akin to the City of London, or at least trying to ape it. Certainly, it has attracted far greater publicity and presents itself much better than any other suburban or provincial town office complex in Britain. Great lengths were taken to make it look a superior development, and no expense was spared in the use of fine quality building materials, fountains, very mature trees, expensive finishing touches, together with good services and public relations. It has, however, been an uphill battle against the establishment of the City of London, who, many would argue, have not graciously accepted the upstarts downstream!

Because of security problems access to Canary Wharf is restricted, including the public viewing area atop the tower (Plate 8.2). (Indeed, panoramic views of London from its few high buildings have been a constant irritant; the long climb to the top of St. Paul's Cathedral is perhaps the best that London can offer.) Compared to Tobacco Wharf, retail development and eating and drinking establishments have been more successful, reflecting a growing office population.

(From Canary Wharf take a DLR train pass South Quay to Crossharbour Station, from East Ferry Road take Crossharbour over the Millwall Inner Dock, cross Millharbour into Tiller Road and turn left on West Ferry Road)

South Quay-Crossharbour-Millwall

This area is markedly different from Canary Wharf and contains a mix of activities. Around the Millwall Inner Dock (also on Heron Quay) are some of the earliest of the Dockland redevelopment in the form of small sheds for business operations, many of them taking advantage of the Enterprise Zone policies. Meanwhile, South Quay has seen the development of a number of high-rise office complexes with various services, including a small retail plaza. Other large land users include the Daily Telegraph works, to the south of Tiller Road, and an Asda superstore to the east of the DLR at Crossharbour.

Millharbour marks the boundary between the new residential developments in the old Millwall Dock to the east and the old working-class dock community of Millwall to the west. The latter is an area of great diversity between the few 19th century terrace houses (seen, for example, on Mellish St.), pre-war and post-war blocks of apartments along Tiller Road (many in a sad state of repair, some vacant, and some

Plate 8.2 Looking east from Canary Wharf, showing the condo developments on the east side of the Isle of Dogs (foreground), the derelict industrial lands of North Greenwich (in the bend of the River Thames) and the derelict Royal Docks (background) (Photo: H.J. Gayler)

destined to be demolished), some immediate post-war two-storey public housing, and a very recent, high density, two-storey, private housing complex off Tiller Road in Claire Place.

(From Tiller Road take a no. D7 bus southbound along West Ferry Road and Manchester Road to Seyssel Street. Cross Manchester Road turn right on Seyssel Street and right on Saunders Ness Road to Island Gardens)

Cubitt Town

Until the opening of the DLR this community at the southern end of the Isle of Dogs was one of inner-London's most isolated communities. The name originates from the 17th century when an English king in residence at the Palace of Greenwich used to exercise his dogs on the marshy flats on the north bank of the Thames. A community here (the centre of which is the intersection of Manchester

Road and Glenaffric Ave.) dates from the time of the opening of the Millwall Docks in 1868 and the development of various industries, warehouses and utilities on the riverside. A railway line was opened from Poplar (to the north of the West India Dock) to Cubitt Town, providing through trains to Fenchurch St. Station in central London, but this closed in the 1920s. Parts of the line, including the viaduct north of Island Gardens Station, were eventually used by the DLR. There was never a through bus service to central London, except by first walking through the foot-tunnel to Greenwich. The isolation of the community was once enhanced by the various entrances and lock gates to the former Millwall and West India Docks, resulting in road closures at frequent intervals.

Most of the early terrace housing in the area has gone (the George Green School, for example, replaced a blitzed section in the 1970s). Newer semi-detached and terrace housing was built after the First World War, and apartment blocks were built in the 1950s and 1960s; much of this has been upgraded since the 1980s. New local shops were part of the post-1945 redevelopment, but recent shopping improvements elsewhere (for example, a direct bus to the Asda superstore) have resulted in a decline in local shopping opportunities. An unusual feature of the area perhaps is the extensive open space that has always existed between the Millwall Dock and the centre of Cubitt Town; part of the area is now a park and the Mudchute Urban Farm.

Between Westferry and Manchester Roads and the River Thames was once a strip of industrial and commercial uses. These are virtually all gone, the buildings demolished, and high-priced town-house and apartment complexes developed in their place. In the rush to maximize private gain, the opportunity was missed to improve public access to the Thames. There is no continuous riverside walkway and few public parks to allow one to view the river scene. Island Gardens itself, close by the DLR terminal, is almost a well-kept secret; it affords a superb view of the many historic buildings in Greenwich on the south bank.

(From Island Gardens take the foot-tunnel under the River Thames to Greenwich. From the River Thames waterfront take King William Walk into Greenwich Park and follow the path up the hill to the Old Observatory. Return to the waterfront)

Greenwich

Greenwich was never regarded as a dockland community, although there was a large population who lived there and could walk to work via the foot-tunnel. It was not included in the LDDC area as there were no extensive areas awaiting redevelopment, although the riverside sites are converting from industrial-commercial to residential in the same way as elsewhere.

Greenwich is a historic centre, predating the wave of 18th and 19th-century development that came downstream. The attraction of the riverside site, close to Blackheath and away from the stench of London, brought royalty here in the 16th century. As royalty deserted Greenwich for country retreats elsewhere, the buildings were taken over by a number of other functions, including the National Maritime Museum, the Royal Naval College and Seamen's Hospital. The present buildings, which include these functions, were designed and built after 1600 by various eminent architects, including Inigo Jones, John Webb, Sir Christopher Wren, Sir John Vanburgh and Nicholas Hawksmoor.

A mixed community and a range of housing was built in the Georgian period that can still be seen today. Deptford, the next community upstream, and Greenwich were to acquire various naval functions, including shipbuilding and outfitting, and the importance of these communities resulted in their attracting London's first railway in the 1830s. The London and Greenwich Railway from London Bridge Station to Greenwich was also the longest set of railway arches ever, built over the poorly drained marshes along the River Thames.

On top of the hill behind Greenwich is the Royal Observatory, founded in 1675 by Charles II, with buildings designed by Sir Christopher Wren. The supremacy of Britain in this field resulted in Greenwich being chosen in 1884 as the dividing line between the western and eastern hemispheres at 0° longitude. The Observatory, which has since moved with a museum taking its place, is located in one of the extensive areas of London parkland; Greenwich Park, dating from the 1400s and formally laid out after 1660, also affords great views of the Docklands and central London.

Greenwich is one of the few tourist areas east of central London. In addition to the National Maritime Museum and Greenwich Park is the Cutty Sark, a former tea clipper between England and South Asia, now in a permanent dry-dock near the foot-tunnel entrance. The nearby streets of the old town centre feature various specialty goods, particularly antiques. Greenwich's tourist traditions are associated with pleasure boat traffic on the River Thames; the advent of the steamboat in the 19th century saw relatively fast and cheap travel between London's East End and the resorts of the River Thames and the Essex and Kent coasts. Greenwich was the nearest and cheapest resort. Today, it is the only place east of central London with regular daily services, at least in summer; the greater convenience of rail, bus and car transportation has virtually eclipsed river travel.

(From Greenwich Pier return to central London on a pleasure boat to Tower Pier; for a faster journey, take a British Rail train from Greenwich to London Bridge and Charing Cross Stations)

River Thames

The return journey from Greenwich to central London by one of the pleasure boat services provides a further perspective on Docklands and the various riverside activities. One will be constantly reminded of the lack of river traffic and the almost eerie calm about the trip; with no commercial docks and few riverside wharves and industry, there is virtually nothing for shipping to serve any longer. Instead, there is a chance to relax and enjoy the ever changing vista on land: the mix of new condominium developments and converted warehouses, the various historic buildings in the older riverside communities of Wapping and Rotherhithe (especially the many pubs!), the massive complex of Canary Wharf, the considerable tracts of derelict land where once stood industry and warehousing, and finally approaching Tower Pier a panorama of famous sights such as Tower Bridge, the Tower of London and St. Paul's Cathedral.

(From Tower Pier transfer to Tower Hill Underground Station)

Further reading

Brownill, S. 1990. *Developing London's Dockland.* London: Paul Chapman.

Coupland, A. 1992. Docklands: Dream or Disaster. In Thornley, A. ed. *The Crisis of London.* London: Routledge, 149-162.

Docklands Consultative Committee (DCC). 1990. *The Docklands Experiment: a critical review of eight years of the London Docklands Development Corporation.* London: DCC.

Chapter 9

West and North-West London: Inter-war Suburbs

In the period between 1850 and 1920 the population of London more than tripled from 2 million to 6.5 million people, but the built-up area only expanded from approximately three miles from the centre to five to eight miles. London was physically a very contained area in this period, the result of a heavy dependence on public transportation by steam railway, the first Underground lines and electric tram.

In the inter-war period, between 1918 and 1939, this compact city was to change for ever. The population increased by 30 percent from 6.5 million to 8.5 million. However, the built-up area expanded by approximately 200 percent, to a distance of about 15 miles from the centre. This impressive landscape change was the result of changing economic, social and technological forces which were unleashed on a system almost entirely devoid of planning. The outcry against this urban sprawl, the waste of good agricultural land, and the jokes about a boring, largely semi-detached, housing environment led to very important changes in British planning legislation and the nature of urban growth in London after 1945.

Our excursion today will focus on this outer suburban area and the forces which brought about change after the First World War. The examples to be used are the west and north-west suburbs of London, part of which was known (and often lampooned) as Metroland, after the Metropolitan Railway, the major developer. We will examine the new type of residential suburb that was developed, including its various services, as well as the expansion of London's manufacturing base.

Socio-economic change and private housing

In spite of the greatest depression the world has ever experienced and the worst slump in the manufacturing sector so far, these were not the worst of times for Greater London. The growth of the service sector was of benefit to London with its diverse economy at an increasingly global scale. Also, manufacturing decline was disproportionately concentrated in the heavy industrial sector which was weak in London. London's diverse industrial base and large number of small units were focused on specialty trades, consumer-based products and the retail sector and global import-export business. Moreover, it was able to take advantage of the many new consumer products entering the market, including plastics and other oil-based products, electrical goods, and car and other road transportation products. Unemployment was lower in the London area than in other parts of Britain. In fact London's industrial and commercial base was attractive enough for there to be considerable in-migration of people, especially to the new industrial areas of west and north-west London.

Unemployment depressed wages, but they were significant enough and sufficiently stable to encourage people to enter the private housing market for the first time. The background to homeownership was similar to the developments in the public sector by the LCC (which were seen at Dagenham in Chapter 7 above): better homes in better surroundings (homes fit for heroes, as the advertising roughly went). Encouraging people to buy their own homes was a means of promoting an improved social status., and the class differences between public and private housing were to become even sharper. Housing for the first time was relatively cheap (although it was still beyond the financial means of many people, even those in employment); large amounts of cheap land were available (in part a hangover of the agricultural depression of the late 1800s), and the housing industry could offer a cheap product because wages and building supply costs were depressed. Meanwhile, financial institutions opened up mortgage borrowing (for long periods at low interest rates) to a wider sector of the population. The lower-middle and skilled working classes, whose parents would have lived in rented accommodation in the older parts of London, moved into the new suburbs 8-15 miles from central London.

Changing transportation provision

The supply of cheap land was promoted, directly or indirectly, by railway companies and to a lesser extent the providers of public road transport; this was in much the same way as for the acres of by-law housing had been developed in the years before the First World War. The promotion was almost entirely by the private sector, and much of the enterprise was with American capital and experience. There were two aspects to the transportation change:

The expansion of the London Underground

Between the late 1890s and 1933 most of the Underground lines and the bus and tram routes became a private monopoly of the Underground Electric Railways of London Limited. Before 1920 there had been limited Underground rail expansion because of high building costs and rival tram companies providing fairly speedy service. However, after 1920 there were long extensions of all the Underground lines up to 15 miles from central London; but with few exceptions these developments preceded the houses, and the Underground lines were in fact overground, sometimes parallel to the main-line railways and providing their suburban services.

These were the Golden Years of the London Underground, although its importance to the capital and various economic and political difficulties resulted in all Underground, tram, trolleybus and bus services coming under public ownership as London Transport in 1933. By 1939 the transport system has reached its limits, both spatially and technologically. The Underground was an (almost) all-electric system of trains stopping at all stations frequently distributed through the new suburbs, with bus feeder services at many of the stations. This imposed a limit of how far suburbs could expand without burdening commuters with too long a journey (45 minutes was the maximum thought possible). Longer journeys would require limited-stopping trains which could not be integrated with stopping trains unless the lines were quadrupled (This has been done to some extent between the Metropolitan/Jubilee lines and Bakerloo/Euston suburban lines in north-west London, the Piccadilly/District lines in west London and the District/Fenchurch St. suburban lines in east London).

Apart from the wartime interruptions to line extensions, the network was finished in 1939. Although the Victoria Line was completed in the late 1960s, the technology was essentially pre-war - a short line with trains stopping at all the stations! The Docklands Light Railway was a radical departure, but an underestimate of need which the future expansion to the traditional Jubilee Line is set to overcome.

British Rail suburban developments

Again these were commenced before the First World War, but came into their own after 1920. The aggressiveness and enterprise of the managers of the Underground were matched by two of the main-line railway companies. (In 1923 the multitude of railway companies in Britain were reduced to four, all of which fanned out from London.) The Southern Railway, which served London's southern suburbs, matched the Underground (and restricted its advance south of the Thames) with its frequent electric rail services stopping at all stations, although it had the advantage of multiple lines and could run slow, semi-fast and fast services over distances that varied from five miles in the inner suburbs to the south coast towns of Brighton (50 miles) and Portsmouth (70 miles). Long-distance electric-train commuting was well established for the middle and upper-income groups before the Second World War.

The London & North Eastern Railway (LNER) attempted the same thing on its lines out of Liverpool Street, following the lead of the former Great Eastern Railway, to the expanding suburbs of east and north-east London. It was hampered by faltering finances, and the electrification of its lines was not completed until after nationalization, in the period 1948-1965. (One of the LNER lines, from Stratford to Epping and Ongar, was handed over to London Transport in the 1940s and became the easterly extension of the Central Line.) In spite of the handicap the LNER, and later British Rail, ran the most efficient steam train suburban service in the world until the 1960s when electrification was completed. Considering that these inner-suburban services stopped at all stations, and that steam trains are not as efficient as electric trains, it was an amazing feat of steam technology and train drivers' abilities that a frequent commuter service could be maintained!

The other British Rail suburban services out of Kings Cross, St. Pancras, Euston, Marylebone and Paddington were cut back as the Underground services expanded and were consequently focused on the outer-suburbs beyond the 12-15 mile limit. Many of these did not come into their own until the post-1960's development of long-distance commuter services.

Suburban sprawl

Social and economic forces and improved transportation were to open up a very different urban landscape after 1920. We have already examined one aspect of this in the public sector housing of the LCC's Becontree Estate (see Chapter 7 above). It was perhaps personified, institutionalized and vilified in the speculative, private housing estates, and especially their semi-detached houses, that were built in the area up to 15 miles from central London.

Tudor Walters Report

The background and driving force behind this was the Tudor Walters Report in 1918. His Committee on Housing included Raymond Unwin who was influential in the designs for the first Garden City at Letchworth and for Hampstead Garden Suburb. The Report stressed the need for public-sector housing, but that private speculative housing would have a role; development would take place on cheap undeveloped land on the fringe of the city, closely tied to public transportation; densities would be a maximum of 12 houses per acre, each designed with a garden front and back; and housing estates plans should be produced by architects and approved by government. Minimum distances between houses of 70 ft. to maximize sunlight, short terraces of houses, recreational spaces, and cul-de-sacs for safety were all part of the design.

It was a far cry from by-law housing, as Dagenham also showed, and was readily accepted by the government of the day, who feared the consequences of not doing something about the poor state of housing in Britain. In spite of 40 years of by-law housing there was still a long legacy of slum housing that would not be finally cleared up for another 40 years! The Ministry of Health became responsible for housing matters and concluded that the slum housing problem could be attacked in two ways: to go up or to go out. The first was thought to be unsatisfactory because of the needs of child care, the problems that could be caused by higher densities, and the lack of a private garden. A self-contained, two-storey house on the urban fringe was thought the better solution - a sentiment, however, that was quietly sidelined in the immediate post-1945 redevelopment of British cities.

What was built was hardly Garden City quality. Whether local councils (being pressured to ease up on spending public monies) or speculative builders (trying to cut corners and increase profits), so many of the housing estates that arose in the 1920s and 1930s followed the Housing Manual to the letter, and the result was the tedious landscape of housing that can be so readily seen and that became the butt of jokes (Plates 9.1-9.2). There was little that was individual about houses; cheap private housing came from pattern books which builders could

Plate 9.1 View of inter-war private housing (Photo: H.J. Gayler)

Plate 9.2 The inter-war (and pre-car) 'semi' (Photo: J.N. Jackson)

handle and architects were long since removed from the scene. Public housing came similarly from the local architect's department.

There was little control over where the housing was built. It was often haphazard, as land became available and by different builders; basic services were necessary for health reasons but how far it was to the shops, schools and various other services was of little concern to anyone but the house buyer. Housing estate aesthetics depended on the cost of the housing and which status group it was designed for: the cheaper the housing, the meaner, barer and more uniform the estate, and the lower the social class group purchasing the housing. Too much housing was built without regard to the quality of agricultural land; and too much was built alongside arterial roads. (Sometimes these were the very by-passes being built around congested town centres, such as the Eastern Avenue around Ilford, resulting in the need for a by-pass around the by-pass 40 years later!)

Semi-detached housing

While this inter-war suburbia features many types of housing design, the most popular by far was the semi-detached house. Before the First World War the semi-detached house, especially if it had a bay window, meant that you had arrived in the respectable middle class. A detached house, or villa, was a step or more beyond. Thus, after 1920, the semi-detached house, however, cheap, gave the appearance of the white-collar, middle class at whatever level; and this contrasted with so much public sector housing (roughly one third of all houses built at this time) which focused on terraces of four, six or eight houses. The abiding fear amongst the new middle class was that their private semi-detached house would be confused with a public one! However, the desire of local councils to save money and for private builders to be able sell status usually resulted in there being little chance of visual confusion. Also, housing of both types was spatially segregated; moreover, the area would be named and clearly identifiable to local residents. The estate's public spaces, the various front gardens, even the house and its name (the middle classes would substitute a name for a street number) would convey a pseudo rural air, again in marked contrast to the earlier by-law housing areas.

While densities were 12 houses per acre or less, estate designs did contain a number of built-in problems. In order to reduce servicing costs, lots were very narrow (25-35 ft.) and as much as 200 ft. long. Only one in ten families had a car by 1939; but that ratio was lower on private estates, and it was quickly realized that lots should be wider in order that a garage could be included. However, the damage was done, and even today many residents either park on the street, if they can, or on the pavement or in the front garden. The move to two-car households brings the problem perhaps to crisis point.

House design presents little variation. The favourites were Mock Tudor with the emphasis on wood beams between the brick or plaster, or neo-Classical with different styles of pillar introduced about front doors or porches. Windows were no longer sash type but opened outwards on hinges, often with leaded panes and containing stain-glass motifs; and at the front of the house there were invariably bay windows (the larger houses having them both up and down). Internal arrangement was fairly standard: hall leading to front sitting room, back dining room and kitchen, stairs leading to two double-bedrooms over living/dining, bathroom over the kitchen, third single-bedroom (often known as the box-room) over the hall. More expensive homes had wider lots with attached garages, french doors to the garden, separate bathrooms and toilets and a downstairs cloakroom, and larger room sizes. Fireplaces were provided since coal was cheap and plentiful; central heating was rarely used for domestic heating until the 1960s in Britain. All houses had a perimeter fence, wall or hedge.

The inter-war community

The rapidity with which these housing estates opened up, the lack of good planning practice until after the 1947 Town and Country Planning Act, and the inability sometimes of local councils to cope with the growth resulted in community services frequently being inadequate for the people who moved there. If the LCC planning whole estates such as Becontree could not get it right, what hope was there for competing builders and hard-pressed local councils. A number of aspects to these communities are considered:

The shopping parade

The commercial centres for these communities had almost the same rubber stamp as the houses. A few would be expansions of an existing town centre, a 19th-century growth point, perhaps a medieval town or village (such as Ilford or Romford). For the most part though nothing existed beyond farmers' fields and a few houses. The focal point for the new community was the railway or Underground station at an emerging major road junction, and extending out from there would be a parade of shops on either side of the road, containing mostly convenience goods and services. Since higher-order goods in any abundance could be some distance away, many of these local shopping parades would contain some of the basic items. Above the shops would be two or more storeys of apartments, often the only apartments in the area. The latest form of entertainment, the cinema, would stand out in the parade, also a new, and very large, pub. Beyond the shops would be larger land uses such as a garage, a park or a school. Because car ownership was so low these shopping parades had virtually no parking spaces, something that would cause undue problems by the 1960s.

Emerging local government

A series of small communities strung out along a railway or Underground line does not make a town or city, a factor already seen in the more compact 19th-century suburban developments. This question of the lack of an urban hierarchy in suburban London came to the fore in the development of higher-order shopping and the reorganization of local government: where is the centre for a number of these small railside communities? Was London to continue being a collection of urban villages spreading out from the City and the West End? Even before 1965 the smaller local government units were an amalgamation of many of these villages; few had recognizable centres, many lay along more than one radial route out from central London.

After 1965 the new 32 London boroughs were often controversial marriages of convenience between a number of very different, and competing communities. Sometimes it was even difficult to agree on a name. The London Borough of Barking was an amalgamation of two working-class boroughs (Barking and Dagenham); the latter resented the name change, or loss, and later the borough was renamed the London Borough of Barking and Dagenham. As a compromise romantic and historic place names were chosen, dating back to Saxon times (9th and 10th centuries), such as Tower Hamlets, Waltham Forest and Havering. There was a definite attempt to go up-market, such as choosing Greenwich over Woolwich, and Hillingdon over Uxbridge.

In the absence of the Greater London Council, which was abolished in 1986, the London boroughs have taken on greater responsibilities. There is no doubt that over the nearly 30 years since they were established they have taken on a greater identity, sense of purpose and political and spatial reality; however, there may not be a dominant centre and there may still be a tug of war between the different areas.

The dormitory suburb function

If the public housing authorities could not do a better job of integrating housing and other activities in the way prescribed by the

Garden City Movement, what hope was there for an array of speculative builders opening up housing estates around railway and Underground stations. There was certainly little in the way of incentive to do otherwise from most local councils. The inter-war dormitory suburb was very definitely sold on the fact that it was a residential-only environment with basic community services in a country setting without the horrors of the industrial city close by. A frequent, cheap train service, or local bus service, was designed to get people to work in a factory or office.

The disadvantages of home-work separation are something that have never been very vigorously pursued. In our planning of the integrated community (for example, the Garden City or New Town) we have recognized the need for a multi-functional city; on the other hand, we constantly improve transportation facilities and subsidize commuter fares, thereby encouraging people to travel longer distances, spending more money and wasting longer periods of time. The inter-war suburban location was now involving people in as much as 2.5 hrs. of travel, door to door, each day to a workplace in central London. Travel delays (which have become notorious in recent years) and too many changes of transport can push this figure even higher. Time spent travelling takes away from family and community life and can be tiring and stressful. Also, catering to peak-hour travel raises public expenditures and wastes resources. (Planning for post-war London has further exacerbated the journey-to-work.)

The Industrial Estate

Although not specifically developed as part of the inter-war suburb, London's many new industries at this time sought locations at the edge of the built-up area. Some of these industries extended older industrial areas, such as the Lea Valley in north-east London or Lower Thameside (for example, the Ford Works at Dagenham and the oil refineries and cement and paper factories further downstream). They depended still on rail and water transportation. However, there was an increasing emphasis on highway-oriented industry, and in addition to various arterial road locations, private entrepreneurs were developing large, well-serviced estates.

(From central London at Baker Street Station take an Amersham train on the Metropolitan Line to Northwood Station)

Metroland

This was the popular name given to an extensive area of north-west London (presently in the London Boroughs of Brent, Harrow and Hillingdon and in Buckinghamshire beyond) that was opened up by the Metropolitan Railway (the forerunner of the Metropolitan Line) and its associated building company in the inter-war period. The development was accompanied by considerable life-style advertising, showing the types of home that could be purchased, their quasi-rural settings, happy children playing outside, mother preparing the tea, the juxtaposition (shown in the slightly oblique aerial view) of home, shops and station with father walking home from the Underground train, and the exhortation of low fares, frequent trains and speedy connections to central London. Variations on the same theme appeared for other railway companies and large building firms.

Our excursion today begins at Baker Street Station which was the location of the former head office of the Metropolitan Railway and is still the principal terminal for Metropolitan Line trains (Figures 9.1-9.2). Note the attempt to speed up the longer-distance Metropolitan Line trains by limiting stops to Finchley Road and Wembley Park, while Jubilee Line trains to Stanmore on adjacent tracks stop at all stations. Further out the British Rail service to Marylebone takes over as the faster service.

The first of the extensive inter-war suburbs, and the first large Metroland estates, are reached at Wembley. A 100,000-seat stadium and exhibition centre were built here in the 1920s, otherwise there is little to distinguish Wembley from surrounding suburbs. The next Metroland estates were built in Harrow, a small, historic Middlesex town, originally built on the hill to the south of Harrow-on-the-Hill Station and dominated by one of Britain's major private schools. As the town grew in the inter-war period as a suburb, a larger commercial centre developed around the station.

(From Northwood Station turn right on Green Lane, cross the road and take no.282 bus through Northwood Hills to Eastcote Village at Joel Street-Eastcote Road junction.)

Eastcote

Our bus trip from Northwood Station to Eastcote takes us through a range of inter-war suburbs. This area can be thought of as part of London's 'stockbroker belt' (which extends northwards into Buckinghamshire) with more expensive, detached housing and even

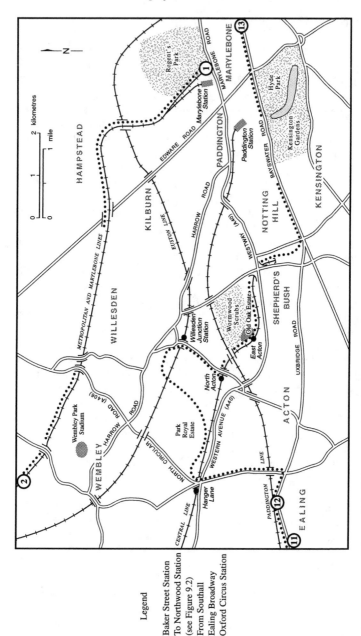

Figure 9.1 West and North-West London Excursion

Legend

1   Baker Street Station
2   To Northwood Station
    (see Figure 9.2)
11 From Southall
12 Ealing Broadway
13 Oxford Circus Station

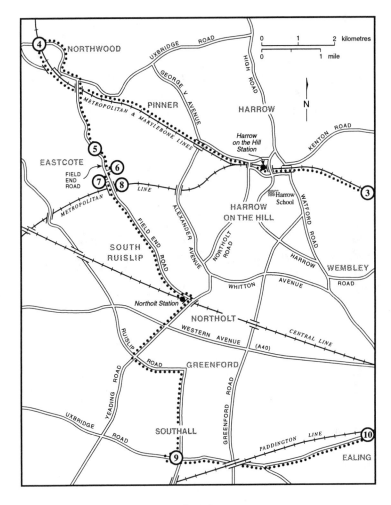

Legend

3   From Baker Street Station
4   Northwood Station
5   Eastcote Village
6   Abbotsbury Gardens
7   Morford Way
8   Eastcote Station
9   Southall High Street/Broadway
10 To Ealing Broadway

Figure 9.2 West and North-West London Excursion (cont.)

lower residential densities. Around the town centre of Northwood the development predates the inter-war period, reflecting the small-scale, early rail commuting by the middle class from towns and villages alongside the main-line railways into London.

The start of the walk is in the old Eastcote Village, where a few of the early buildings remain, and where the London Borough of Hillingdon has promoted a walking tour for those interested in the history of the area.

(Walk ahead along Field End Road, turn left at Abbotsbury Gardens, turnaround and left into Field End Road, right on Morford Way, left on Hawthorne Avenue, left on Elm Avenue, right on Field End Road to Eastcote Station)

Eastcote itself developed further south around the station on the Metropolitan Line branch to Uxbridge. The branch was opened just before the First World War, but the major period of development was delayed until the 1920s and 1930s. We will examine a number of features about the inter-war suburb outlined above:

The shopping parade

To the north of the station along Field End Road is a very typical parade of shops, with apartments above, that accompanied the opening of the station and surrounding residential areas. The convenience and some higher-order goods in part meet the needs of local residents. But the shops themselves are small and cannot be adapted to present-day needs, especially the move to cheaper, superstore and other retail warehouse shopping. Furthermore, these once pleasant shopping environments have not adapted well to the car, let alone the large amounts of parking that larger shopping units would generate. Field End Road at least has service roads and parking immediately in front of the shops (many inter-war shopping parades on busy roads are not as fortunate); but this is congested, necessitating car-park development behind the shops, as well as parking and increased traffic spilling over on to residential streets.

The speculative housing estate

Abbotsbury Gardens typifies this semi-detached speculative development of the 1920s-1930s: long lines of houses constructed by the same builder and at the time varying hardly at all. Developments like these were to draw very mixed reactions. Osbert Lancaster, the cartoonist, was scathing in his criticism of this suburban sprawl. John Betjeman, the poet, on the other hand, waxed eloquently about how the country had come to the suburbs.

For all the criticism of this semi-detached sprawl, the basic house had stood the test of time very well. Unlike by-law housing, it has never been demolished (except by enemy bombing or if in the way of some redevelopment project), but continually upgraded and frequently extended as succeeding generations required greater comforts and more space. So much so, that for people interested in architectural heritage, it is exceedingly difficult to find an original inter-war house (or pair of houses). Car parking may be the only pressing physical problem in such areas.

Conserving inter-war development

The London Borough of Hillingdon has been responsible for designating certain parts of the borough as Conservation Areas in order to protect a significant type of development. One of these is Morford Way/Morford Close. These streets were laid out after 1918 and contrast with the earlier, more speculative and uncontrolled development on the nearby "treed" streets (e.g. Elm Avenue). The architect, Frank Osler, who had already been responsible for houses in Hampstead Garden Suburb, was engaged by the developer; and thus the street has special architectural and historical significance. Conservation in this way does affect individual property rights, in particular the types of change that can be undertaken; although the irritants and higher costs involved are frequently matched by relatively higher property values compared to surrounding areas.

Eastcote Station

A feature of corporate control in London Transport, both before and after nationalization in 1933, was the unifying image and uniformity in design and management that was so well projected by Lord Ashfield and Frank Pick, respectively chairman and commercial manager (and something that most other railway companies did not match). Examples include the logo, colour use, station design, train and bus models, and the geometric map of the London Underground. Much of this image was the responsibility of Charles Holden, who was the chief architect and who designed simple, modernist, yet architecturally significant, stations along the new Underground lines. Eastcote Station is a smaller example of this style.

(From Eastcote Station take a no. 282 bus to Northolt Station, change to a no. 120 bus to Southall High Street. Walk along part of the High Street and Broadway)

Southall

This was once a very unspectacular late 19th and early 20th-century suburb, straddling the Uxbridge Road and the Great Western Railway from Paddington Station. It was once a separate local authority within Middlesex, and since 1965 it has been part of the London Borough of Ealing. More recently, Southall has been identified as having one of the larger south Asian population concentrations in London. In 1981 there were some 190,000 people of Indian, Pakistani and Bangladeshi origin in Greater London. At that time, Ealing had 25 percent of its population recorded as coloured immigrant; only Brent with 33 percent had a higher concentration. However, both here and elsewhere in London there is little ghettoization. Two wards in Ealing have 85.4 and 71.0 percent of their populations as coloured immigrants, but these are quite exceptional.

The Asian population is reasonably dispersed across the metropolitan area, and if it concentrates at all, it is in the outer rather than the inner suburbs. This is in part because Southall and nearby Hounslow, where there is also an Asian concentration, are close to Heathrow Airport, the major entry point for these people. Furthermore, the Airport area is a major employment base, demanding a range of skills that a new immigrant group could take advantage of.

The Asian population has remained relatively prosperous, especially in west London, and has made its mark in business, particularly the retail sector, catering to its own group and taking over corner-stores throughout London. More modest demands allow them to survive economically; also, they provide a service that others are not prepared to duplicate. Southall is very much an Indian, especially Sikh, suburb; and its cultures, religions, dress, and languages stand out from the host population. There is little attempt at assimilation; and economically the group has established itself well, contrasting quite markedly with the Bangladeshi population seen in Chapter 6 above in Whitechapel area. As a result, there has been considerable resentment and some racist attacks against Asians.

Southall Broadway and High Street over the last 30 years has become almost entirely a retail and service centre for this Sikh population. The various functions and types of product identify with different aspects of that culture, and service can be provided in local languages.

(From Southall High Street take a no. 207 or 607 bus to Ealing Broadway)

Ealing

This is one of London's late Victorian, middle-class suburbs; although what has frequently happened is that large homes have been subdivided into apartments, resulting in a higher proportion of single person households and lower income people. Where the older homes were in spacious grounds, their place has sometimes been taken by apartment blocks.

Ealing Broadway is one of inner-London's large regional centres and is similar to that seen in Chapter 7 above in Ilford. The traditional shopping street has been augmented with off-street shopping malls. However, the concentration of shopping here, the focus of bus, rail and Underground routes, and the high car use in a middle-class area have caused considerable congestion and loss of environmental quality.

(From Ealing Broadway take a PR1 bus through Park Royal Trading Estate to Willesden Junction Station)

Park Royal

This industrial estate in West London opened just beyond the built-up area in the late 1920s. It is the largest estate in Greater London, and at its height in the 1960s employed almost 40,000 people. It is well served by main-line railway, Underground and road transport; however, after the 1960s it was heavily affected by the declining importance of manufacturing in London. The bus trip through this area will show that a number of non-manufacturing activities have subsequently taken over, including superstore and retail warehouse shopping, sports facilities, hospital, and other service industries.

(From Willesden Junction Station take a 266 bus to North Acton Station, change to a Central Line train to East Acton Station. Turn left on Erconwald Street, right on Braybrook Street, right on Wulfstan Street, left on Erconwald Street to East Acton Station)

LCC's Old Oak Estate

This housing development between East Acton Station and Wormwood Scrubs stands in marked contrast to what we have seen so far. It is one of the early attempts by the LCC to build a small satellite estate of some 300 homes on the edge of the built-up area, a forerunner of the post-1920 out-county estate seen in Chapter 7 above at Dagenham. It was commenced just before the First World War but did not come into its own until the Central Line was extended through the area in 1920. The estate has Wormwood Scrubs as its large open space (and Her Majesty's Prison!) separating the area from older and more crowded industrial and residential developments. The curved streets (contrasting with the grid-iron pattern of by-law housing), gable ends and country cottage appearance give the estate a unusually, but distinctly, German vernacular flavour.

(Return to central London from East Acton Station)

Central Line

The journey back to central London on the Central Line, especially in rush-hour, will dramatically point out the chronic underinvestment, overcrowding, poor decision-making, labour shortages and travellers' frustrations concerning transportation in London. The slow journey, perhaps one or more trains being taken out of service, and the wall-to-wall people by the time the train reaches central London point to a need for new lines. The proposed solution is a fast, limited-stop, east-west Underground line (the Crossrail scheme paralleling the Central Line) that would connect the suburban commuter lines out of Liverpool Street and Paddington Stations and provide extensive areas east and west of London with better connections to central area destinations. This is modelled on the RER in Paris where there are already five such lines. However, London is far behind; Thameslink is the only service, connecting the suburban services from St. Pancras with some of those south of the river. Meanwhile, Crossrail was delayed in favour of the Jubilee extension to Docklands.

Postscript: the end of an era for London's growth

Having looked at a large area of London that developed in this inter-war period, it is instructive to look at what became of it. During the 1920s and 1930s people became distressed at the nature of the urban development taking place; eventually the government took action, although it was not until the late 1940s that effective planning and development legislation was in place.

What upset people in the early 1920s was that in spite of the imaginative planning of the Garden City Movement, those in control in central and local government in effect dismissed it. The Movement and the whole of New Town planning was put back 30 years. First, government was worried by the blight that extensive public housing schemes would cause on the countryside and strongly middle-class small town and rural communities. Second, there was the expense of providing fully integrated communities at a time of declining economic fortunes. Third, allowing satellite communities, such as Dagenham in the public sector or Metroland in the private sector, would still satisfy the ideals set out in the Tudor Walters Report and provide for a much better environment than existed in by-law housing or earlier slum housing. Fourth, improved transportation would allow these people in the new communities to commute to their jobs in inner London.

The Garden City Movement and its fellow travellers were later able to say, 'I told you so.' The Dagenhams and Metrolands of this world were to bring a spate of problems which desperately needed attention:-

1) *Population growth* Growing by over 2 million in just under 20 years was thought excessive, and far ahead of other parts of the country. The attraction was London's employment base which again grew at a rate far ahead of the national average. There was concern that without any government intervention a two-nation approach to economic development was emerging in Britain: a prosperous Southern England and Midlands and a depressed Northern England, Scotland, Wales and Northern Ireland. Whilst London, for example, had an unemployment rate under 10 per cent, some industrial communities in the North had rates as high as 50 per cent. There was growing interest in redistributing economic development in order to help depressed communities and stop the shift of population to the London area.

2) *The extent of urban expansion* There was growing concern at the unregulated, low-density nature of the urban expansion taking place, the almost complete disregard for the quality of agricultural land, and the strip-like sprawl along main roads which sterilized even more land. Also, at 12-15 miles from central London, suburban development was reaching the limit of the transport technology to provide commuters with a reasonable journey time to work.

3) *The nature of urban development* Both the LCC and the private speculators were largely housing developers and paid little regard to

more than the basis services. Whilst the local council and other private companies would look after the various utilities, education and medical needs, police and fire prevention, retailing and employment, it was too often a hit-or-miss affair, and services often lagged behind population growth. These suburbs were largely dormitory communities, and there was a growing awareness that this was not satisfactory.

4)   *Urban design characteristics*   Architects, planners, the Council for the Preservation of Rural England, writers, journalists, Colonel Blimp types etc. cried out about the way this sprawl was disfiguring the countryside. Rather than bringing the countryside to the suburb, urban development in its various manifestations was extending out in linear form along main roads obscuring the countryside. Meanwhile, the various types of urban development were often discontinuous and wasted agricultural land.

5)   *The clash of cultures*   There were also objections to the new suburbanites   Cities had been quite contained before 1920; and the countryside beyond was the preserve of the landed gentry and an upper-middle class elite, together with a rural working class who would be largely employed on the farm or in various service occupations and would in no way threaten the natural order. The electric train, bus and car brought about the democratization of the countryside, an invasion of people who were not beholden to the old order. Besides new housing developments, large numbers of people visited the countryside on weekends or for holidays, and a wave of small, ramshackle bungalows (of the type experienced by Basildon) appeared all over southern England on any land where the farmer could not make a decent living.

Various measures were taken to reverse these developments:-

1)   *An Act of Parliament in 1935 to restrict ribbon development.*

2)   *The establishment of the Barlow Commission in 1937* to study population growth and economic development in Britain, hear public submissions and report to the government. This was done in 1940 and the restriction of London's growth was recommended through a system of licencing further industrial expansion and its location. This was legislated in the Distribution of Industry Act, 1945.

3)   *A Green Belt of about five miles in width* established beyond the 1939 built-up area of London and incorporating the various towns and villages found there. It would be an effective barrier to spatially

continuous growth, as well as provide Londoners with access to the countryside. This was recommended in the Barlow Report, and endorsed by Abercrombie's Greater London Plan. Further development of London would be sufficiently far away beyond the Green Belt that more self-contained communities would ideally be encouraged. These Green Belts have been extended rather than contracted; there is vigorous opposition in Britain from rural and small town interests to the loss of Green Belt, something the market economy and Thatcherite policies have so far been unable to break.

4) *Sir Patrick Abercrombie's Greater London Plan* He reported in 1944, recommending the improvement of road and rail communications and the containment of growth partly through the construction of eight government-financed New Towns, varying in size from 30,000-50,000, in the counties around London. The New Towns Act, 1946 evolved from this plan. The LCC's out-county estates were to continue after 1945; and the Town Development Act, 1952 allowed for agreements between councils for the longer distance dispersal of population from London and other large cities. 60 percent of London's growth of one million, however, would be in the private sector and would consist of the expansion of small towns between 30 and 50 miles from London.

5) *Town and Country Planning Act, 1947* This required all planning authorities (the counties and county boroughs) to prepare development plans for government approval, and in the context of these to require planning permission for all development proposals (development control), with the right of appeal over a local decision to central government. The right to develop all land was now nationalized, although land still stayed in private ownership, and the land market would effectively control prices. The county plans resulted in an effective Green Belt around London. Together with other proposals, the emphasis in the plans was on the negative, preventing the largescale overspill development of London. Increasingly, it had to be absorbed into the development proposals of individual towns and cities.

These various measures, along with the outbreak of war in 1939, literally stopped London's growth in its tracks. The Green Belt resulted in the abandonment of Underground rail extensions and numerous housing developments. As we saw east of Upminster and in Plate 7.2 above, this sharp physical break between town and country remains today like a time warp, and London's growth after 1945 had to leap-frog the Green Belt.

Further reading

Bashall, R. and Smith, G. 1992. Jam today: London's transport in crisis. In Thornley, A. ed. *The Crisis of London*. London: Routledge, 37-55.

Hall, P. 1988. *Cities of Tomorrow*. Oxford: Basil Blackwell, 47-85.

Hall, P. 1992. *Urban and Regional Planning*. 3rd. edition. London: Routledge, 12-89.

Jackson, A.A. 1973. *Semi-Detached London: Suburban Development, Life and Transport, 1900-39*. London: Allen and Unwin.

Jones, E. 1991. Race and Ethnicity in London. In Hoggart, K. and Green, D.R. eds. *London: A New Metropolitan Geography*. London: Edward Arnold, 176-190.

Stevenson, J. 1984. *British Society, 1914-45*. The Pelican Social History of Britain. Harmondsworth, UK: Penguin Books, 221-243.

Weightman, G. 1984. *The Making of Modern London, 1914-1939*. London: Sidgwick and Jackson.

Chapter 10

Hampstead

London is made up of a large number of local communities, or what is often referred to as urban villages. In the last hundred years they have been amalgamated to form the 32 London Boroughs that we see today. However, these local communities are still very real in the everyday lives of Londoners; indeed, they often have greater visibility than the borough of which they are a part. Historically, they may vary from a medieval village, long since swallowed up by suburbia, to a late 19th-century, early 20th-century railway suburb, dockland community, industrial suburb or LCC out-county estate. Yet the characteristics common to nearly all of them is a measure of geographical compactness, nodality or focus of routes at some centre, a concentration there of various community services, a degree of homogeneity and common interest among the local population, and an air of distinctness from nearby communities that may be given by form, function or history.

We have already seen a number of these local communities in passing as we have studied various aspects of London's development (for example, Shadwell in the East End or Eastcote in north-west London). This excursion specifically examines a number of these communities in the inner-north London area around Hampstead. The choice is purely personal: I once lived there! The reader could explore any one of a number of such communities, including ones already mentioned.

Hampstead, however, does have a number of features that make it stand out in the minds of Londoners. Its country village-like atmosphere has been well preserved; its location and physical features have promoted its cause; its earlier development was not overwhelmed by 19th and 20th century suburbia; and it attracted a group of people who not only publicized the place but have been keen to protect it. Since 1965 Hampstead has been part of the London Borough of Camden (along with St. Pancras and Holborn), but it has lost little of its distinctiveness and has little in common with areas to the south.

Our excursion will look at the various characteristics of this community, including Hampstead Heath, the village itself, the adjacent communities of Golders Green and Hampstead Garden Suburb, and briefly the community of Highgate on the other side of the heath (Figure 10.1).

(From central London at Trafalgar Square take a no. 24 bus (or from Aldwych take a no. 168 bus) to Hampstead Heath Station. From South End Road turn right on South Hill Park and right on Parliament Hill and through to the heath, turn right to Parliament Hill lookout)

Hampstead Heath

To the north and east of the village is an extensive area of open space which over the years has been protected from urban development, first through large private land holdings and later through the public ownership of land and planning controls. Hampstead Heath is owned and administered by the Corporation of the City of London. The heath, which rises to just over 400 ft. above sea level, is an outlier of the Bagshot Beds of sand and gravel dating from the Tertiary geological period. It was considerably altered during the Pleistocene period by glaciation and the changing course of the River Thames (it once flowed to the north of Hampstead Heath). Between central London and Hampstead Heath the route crosses the Taplow and Boyn Hill terraces of the River Thames.

Hampstead Heath and surrounding higher land acted as a barrier to early rail communications into London from the north (Figure 10.1). It can be seen that in the short distance between Primrose Hill, north of Regent's Park, and the heath there are five lines that jostle for space in the area: the main lines into St. Pancras, Euston and Marylebone Stations, the Metropolitan Line to Baker Street and the North London Line from Richmond and Willesden to Stratford. The branch line from Gospel Oak to Walthamstow and Barking and the former Great Northern branches from Kings Cross Station to Edgware and High Barnet were also forced to skirt the area. It was not until the Underground era at the end of the 19th century that the various branches of the Northern Line could serve the area directly.

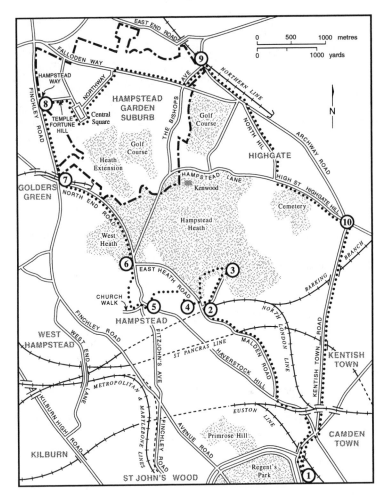

Legend

1	To and from central London	7	Golders Green Station
2	Hampstead Heath Station	8	Hampstead Garden Suburb Gateway
3	Parliament Hill	9	East Finchley Station
4	Keat's Grove	10	Archway Station
5	Hampstead High Street	—·—	Hampstead Garden Suburb
6	Whitestone Pond		Conservation Area Boundary

Figure 10.1 Hampstead Excursion

The heath has become an integral part of London's open space; it plays an important role in the recreational pursuits of both local residents and visitors, creates the feeling of countryside, and influences property values in surrounding areas. The heath has been retained in a fairly natural state, although there are more organized recreational activities on its fringes, including fun-fairs at holiday times. Some dammed-up ponds act as outdoor bathing areas, and the former private mansion and grounds of Kenwood at the northern end of the heath are open to the public. From Parliament Hill there is a panoramic view south over London.

(From Parliament Hill walk west towards Hampstead across the Ponds to South End Road, cross to Keats Grove, right on Devonshire Hill, left on Willow Road, left on Flask Walk, cross Hampstead High Street, left on Heath Street, right on Church Row, right on Holly Walk, left on Holly Hill to Hampstead Hill and Whitestone Pond)

Hampstead village

The place's claim to fame is that it has long attracted the attention of people in the art, architectural and literary worlds. Plaques to famous people who have resided in Hampstead abound, including Keats (whose house on Keats Grove is open to the public), Constable, Gailsworthy, Romney, and Gilbert Scott, and no doubt more will be erected for present residents. The somewhat chaotic medieval nature of the village still comes through on our walk, even though most of the buildings date from the 17th and 18th centuries when the village with its mineral water wells developed as a fashionable spa. The shops along the High Street have modern facades but they mask 18th-century buildings. Specialty shops and stylish restaurants and bars abound on this street, reflecting the upper-middle class character of the neighbourhood. Church Row with its terraces of early 18th century houses is probably associated with the development of the spa. This highly regarded street leads to Hampstead Parish Church which was built in 1747, replacing an earlier, medieval church.

(From Whitestone Pond take a no.268 bus to Golders Green Station)

Golders Green

After Hampstead we cross the old 1889 County of London boundary into the former county of Middlesex, today the London Borough of

Barnet. This area north of the heath was largely rural with small towns and villages until the 1920s. In spite of the main railway line to St. Pancras and the branches of the Kings Cross line traversing the area, there was no extensive suburbanization until the development of the Northern Line of the Underground. It reached Golders Green in 1907 and was extended to Hendon Central in 1923 and Edgware in 1924. The Kings Cross branches from Finsbury Park were taken over at this time by the Northern Line through Kentish Town and Highgate.

Golders Green is largely a look-alike inter-war suburb. Its extensive Underground rail workshops are associated with its terminal status for so many years. The station is also an important interchange point for bus services in north-west London, and a commercial centre has grown up to serve this part of the borough. The Golders Green Hippodrome, opened in 1913, was at one time the largest music hall in London. It has long since become a TV studio; in fact nearly all of London's suburban theatres have closed, as have most of its large cinemas.

After 1920 Golders Green attracted a sizable Jewish population. This was part of a 19th-century migration from the Spitalfields-Whitechapel area (see Chapter 6 above), into places such as St. John's Wood and Hampstead, and thence to Golders Green and later further north into Hendon. The London Borough of Barnet presently has the largest congregation of Jewish people in London (at over 15 percent). Two important factors in this pattern have been the upward movement in socio-economic status and changing location of Jewish cultural and religious institutions; many of these institutions can be seen in the area.

(From Golders Green Station take a no. 82, 102 or 260 bus along Finchley Road (northbound) to Bridge Lane/Temple Fortune Lane junction. Cross Finchley Road to Hampstead Way, left on Temple Fortune Hill, right on Erskine Hill to Central Square, exit on Northway to Falloden Way)

Hampstead Garden Suburb

Misnamed because it was never in Hampstead, but its development embraced extensions northward of Hampstead Heath. In part its development and the heath's extension were to protect the area from inferior speculative development, associated with the extension of the Northern Line and the opening of a new station below ground level between Hampstead and Golders Green. (The station was never opened and remains a ghost station.)

The Garden Suburb is one of the best examples in Britain of 20th-century town planning and domestic architecture, and follows to some degree the Garden City principles of Ebenezer Howard. The character of the area, which includes imaginative architecture and lay-out and overall harmony between the various human and physical elements, resulted in a Conservation Area designation in 1968 by the London Borough of Barnet (Figure 10.1). In 1977, the British government declared the area to be of Outstanding Architectural and Historical Interest. It has become one of London's most prestigious residential areas, although this was not the intent of its founder.

Hampstead Garden Suburb dates from 1906 when Dame Henrietta Barnett, a local resident, was inspirational in setting up the Hampstead Garden Suburb Trust Ltd., which in turn purchased 243 acres of land near the Hampstead Heath Extension from the Trustees of Eton College (the elite boys' private school near Windsor). Barnett was an admirer of the Garden City Movement; and Sir Raymond Unwin, who was already involved in several Garden City type schemes, was appointed architect to the Trust. Plans were drawn up for some 8,000 houses, with a Central Square with churches and an Institute at the highest point in the area. The desire to improve upon by-law restrictions resulted in the passing of the Hampstead Garden Suburb Act in 1906. This allowed roads to be reduced from 40 percent in a typical by-law scheme to 17 percent, and for gardens and open space to increase from 17 to 55 percent. More informal layouts were introduced, including cul-de-sacs and curving streets, and a range of house types reflected the wide socio-economic appeal that the developers intended for the area.

Sir Edwin Lutyens, a major British architect, was responsible for many of the public buildings, and along with other important architects gave the Garden Suburb its outstanding residential quality. The place does have its unusual qualities: the town wall, gatehouses and gateway off the Finchley Road have a distinct medieval German touch about them, contrasting markedly with the English country village housing (Plates 10.1-10.2)., whilst the Central Square, as Hall (1988, 105) puts it, 'looks as if it is waiting for an Imperial Durbar that will never now take place.'

The Garden Suburb, however, did not entirely follow the principles set out by the Garden City Movement: there was no industry, few services and in no way would it be remotely self-contained. The Underground stations allowed for easy commuting, although many residents were early car-owners. Also, there was scant attention paid to the fact that the new housing should be for the relief of people living in squalid conditions. Artisan housing was built, but the reputation that

Plate 10.1 Hampstead Garden Suburb: Finchley Road gateway (Photo: H.J. Gayler)

Plate 10.2 Hampstead Garden Suburb street scene (Photo: H.J. Gayler)

the Garden Suburb quickly attained led to it becoming an upper-middle class residential area.

(Cross Falloden Way and take a no. 102 bus to East Finchley Station and 143 bus to Highgate High Street. Walk along High Street and Highgate Hill to Archway Underground Station)

Highgate

A village on the eastern side of Hampstead Heath, and similar in so many respects to Hampstead, dating from the 18th century, a former spa and a haven for people from the art and literary tradition, and containing many fine terraces and individual houses. Close to the High Street, Highgate Cemetery, dating from the 1830s, has become a tourist feature; guided tours are conducted of its maze of tombs, mausoleums and catacombs, and personal histories of both the famous and the forgotten characters are recalled. The most renown tomb is perhaps that of Karl Marx.

(From Archway Underground Station return to Central London on the Northern Line of the Underground, or 134 or 135 bus to Tottenham Court Road)

Further Reading

Borer, M.C. 1976. *Hampstead and Highgate: The Story of Two Hilltop Villages*. London: W.H. Allen.

Green, B.G. 1977. *Hampstead Garden Suburb, 1907-77*. London: Hampstead Garden Suburb Residents' Association.

Hall, P. 1988. *Cities of Tomorrow*. Oxford: Basil Blackwell, 101-105.

Waterman, S. and Kosmin, B. 1988. Residential patterns and processes: a study of Jews in three London boroughs. *Transactions, Institute of British Geographers* 13: 79-95.

Bibliography

Bashall, R. and Smith, G. 1992. Jam today: London's transport in crisis. In Thornley, A. ed. *The Crisis of London*. London: Routledge, 37-55.

Borer, M.C. 1976. *Hampstead and Highgate: The Story of Two Hilltop Villages*. London: W.H. Allen.

Brownill, S. 1990. *Developing London's Dockland*. London: Paul Chapman.

Brownill, S. and Sharp, C. 1992. London's Housing Crisis. In Thornley, A. ed. *The Crisis of London*. London: Routledge, 10-24.

Clayton, K.M. ed. 1964. *Guide to London Excursions*. London: 20th International Geographical Congress.

Clout, H.D. ed. 1991. *The Times London History Atlas*. London: Times Books.

Clout, H.D. and Wood, P.A. eds. 1986. *London: Problems of Change*. London: Longman.

Coupland, A. 1992. Docklands: Dream or Disaster. In Thornley, A. ed. *The Crisis of London*. London: Routledge, 149-162.

Cross, M. 1992. Race and Ethnicity. In Thornley, A. ed. *The Crisis of London*. London: Routledge, 103-118.

Diamond, D.R. 1991. The City, the 'Big Bang' and Office Development. In Hoggart, K. and Green, D.R. eds. *London: A New Metropolitan Geography*. London: Edward Arnold, 79-94.

Docklands Consultative Committee (DCC). 1990. *The Docklands Experiment: a critical review of eight years of the London Docklands Development Corporation*. London: DCC.

Edwards, M. 1992. A Microcosm: Redevelopment Proposals at Kings Cross. In Thornley, A. ed. *The Crisis of London*. London: Routledge, 163-184.

Gayler, H.J. 1970. Land Speculation and Urban Development: Contrasts in South-East Essex, 1880-1940. *Urban Studies* 7: 21-36.

Green, B.G. 1977. *Hampstead Garden Suburb, 1907-77*. London: Hampstead Garden Suburb Residents' Association.

Hall, P. 1962. *The Industries of London since 1861*. London: Hutchinson.

Hall, P. 1964. General Introduction to the excursions in Central London. In Clayton, K.M. ed. *Guide to London Excursions*. London: 20th International Geographical Congress, 22-28.

Hall, P. 1988. *Cities of Tomorrow*. Oxford: Basil Blackwell.

Hall, P. 1989. *London 2001*. London: Unwin Hyman.

Hall, P. 1992. *Urban and Regional Planning*. 3rd. edition. London: Routledge.

Harrison, P. 1992. *Inside the Inner City: Life under the Cutting Edge*. Harmondsworth, UK: Penguin Books.

Harwood, E. and Saint, A. 1991. *London: Exploring England's Heritage*. London: HMSO.

Jackson, A.A. 1973. *Semi-Detached London: Suburban Development, Life and Transport, 1900-39*. London: Allen and Unwin.

Jones, E. 1991. Race and Ethnicity in London. In Hoggart, K. and Green, D.R. eds. *London: A New Metropolitan Geography*. London: Edward Arnold, 176-190.

Lawrence, H.W. 1993. The Greening of the Squares of London: Transformation of Urban Landscapes and Ideals. *Annals of the Association of American Geographers* 83: 90-118.

Michie, R.C. 1992. *The City of London: continuity and change, 1850-1990*. London: Macmillan.

Morgan, B.S. 1991. The Emerging Retail Structure. In Hoggart, K. and Green, D.R. eds. *London: A New Metropolitan Geography*. London: Edward Arnold, 123-140.

Porter, R. 1994. *London: A Social History*. Cambridge, MA: Harvard University Press.

Pryke, M. 1991. An international city going 'global': spatial change in the City of London. *Environment and Planning D: Society and Space* 9: 197-222.

Punter, J. 1992. Classic Carbuncles and Mean Streets: Contemporary Urban Design and Architecture in Central London. In Thornley, A. ed. *The Crisis of London*. London: Routledge, 69-89.

Rubinstein, A. ed. 1991. *Just Like the Country: memories of London families who settled the new cottage estates, 1919-1939.* London: Age Exchange.

Stevenson, J. 1984. *British Society, 1914-45.* The Pelican Social History of Britain. Harmondsworth, UK: Penguin Books.

Summerson, J. 1962. *Georgian London.* rev. edn. Harmondsworth, UK: Penguin Books.

Waterman, S. and Kosmin, B. 1988. Residential patterns and processes: a study of Jews in three London boroughs. *Transactions, Institute of British Geographers* 13: 79-95.

Weightman, G. 1984. *The Making of Modern London, 1914-1939.* London: Sidgwick and Jackson.

White, H.P. 1971. *London Railway History.* Newton Abbot: David & Charles.

Willmott, P. 1963. *The Evolution of a Community: a study of Dagenham after 40 years.* London: Routledge & Kegan Paul.

Young, M. and Willmott, P. 1962. *Family and Kinship in East London.* Harmondsworth, UK: Penguin Books.

Index

About the Author

Hugh J. Gayler is an Associate Professor of Geography at Brock University, St. Catharines, Ontario. He graduated with a BA from the University of Leicester, an MA from the London School of Economics and a Ph.D from the University of British Columbia. He specializes in urban geography, and has published on various aspects of suburbanization and urban expansion into areas of high resource value. Previous books include *Retail Innovation in Britain: The Problems of Out-of-Town Shopping Centre Development* (1984), and editor and contributor for *Niagara's Changing Landscapes* (1994).